DATE DUE

VET TECH ESSENTIALS
Core Principles in Veterinary Technology

Amy J. Wolff, DVM

PEARSON

Boston   Columbus   Indianapolis   New York   San Francisco   Upper Saddle River
Amsterdam   Cape Town   Dubai   London   Madrid   Milan   Munich   Paris   Montreal   Toronto
Delhi   Mexico City   Sao Paulo   Sydney   Hong Kong   Seoul   Singapore   Taipei   Tokyo

Editorial Director: Vernon Anthony
Senior Acquisitions Editor: William Lawrensen
Program Manager: Alexis Duffy
Editorial Assistant: Lara Dimmick
Director of Marketing: David Gesell
Marketing Manager: Stacey Martinez
Assistant Marketing Manager: Alicia Wozniak
Senior Marketing Assistant: Les Roberts
Senior Managing Editor: JoEllen Gohr
Project Manager: Kris Roach
Operations Specialist: Deidra Skahill
Senior Art Director: Jayne Conte
Cover Art: iStock Photo
Cover Designer: Karen Noferi
Media Project Manager: April Cleland
Full-Service Project Management: Cenveo Publisher Services
Composition: Cenveo Publisher Services
Printer/Binder: Edwards Brothers
Cover Printer: Lehigh-Phoenix Color/Hagerstown
Text Font: Adobe Garamond Pro

Credits and acknowledgments borrowed from other sources and reproduced, with permission, in this textbook appear on the appropriate page within text.

Library of Congress Preassigned Control Number
2013949375

10 9 8 7 6 5 4 3 2 1

PEARSON

ISBN 10: 0-13-508016-9
ISBN 13: 978-0-13-508016-0

Dedication

To the student—and the patients who will be served by
your caring, your compassion, and your commitment.

Brief Contents

Contents

3
Legal and Ethical Issues in Veterinary Medicine   73

4
The Animal Industry   107

5
Developing People Skills and Work Ethics   142

Preface

The opportunity to participate in the development of a program in veterinary technology was presented to me in 2006. After sixteen years of veterinary practice, half of which was spent in an emergency clinic with a talented support staff, I was confident I could design a curriculum and instruct the veterinary technology student in the skills and concepts necessary to provide competent medical and surgical care.

I had spent many years in a college classroom in preparation for my own career, and I was certain I could replicate the learning environment in which I was trained. But today's classroom is filled with diverse learners who have become accustomed to distinctly different ways of obtaining and processing information. My students appeared reluctant to read, especially long pages of text. Accustomed to obtaining information quickly over the Internet, they often lacked the ability to critically evaluate content and source. I wanted to produce educational material that would inform and engage the student in a concise manner, written in active voice.

My goal was to write an introductory book for the entry-level veterinary assistant and technology student. The majority of applicants seeking entrance into the field of veterinary assisting and veterinary technology express a deep-seated connection to animals, often preferring their company to that of people. When the prospective student is asked if they have any experience *working* with animals, the most common response involves lifelong animal ownership or caring for large numbers of a single species. For the student with no working knowledge of veterinary medicine as a career, this area of study is an educational awakening. Having looked no further than the love they harbor for animals, they are often surprised when they learn about veterinary medicine as a business, filled with regulations, governing bodies, ethical dilemmas, and legal encumbrances. They begin to understand that there is a vast difference between people who *love* animals and people who *work* with animals.

The chapter subjects of this book were selected based on their relevancy for beginning students. Recognition of the contributions of the veterinary profession to maintaining public health and a safe food supply was a primary objective. Too often the student is unaware of the importance of the profession in these vital functions. Legal, ethical, and business components are chapter topics, guiding the student in understanding the challenges of practice and how veterinary medicine is governed at the local, state, and federal levels. Lastly, a section on personal development and work ethics makes the student aware of the "soft skills" that must be developed to maintain good workplace relationships and become a valued employee.

The book contains supplemental material for instructors to use in part or whole. Having spent countless hours writing curriculum and preparing lectures, PowerPoint presentations, and exams, I wanted the new instructor to have a platform of ideas from which to work and develop their own unique content. Ideas for classroom activities and discussions can be modified to fit the instructor's classroom environment and the specific needs of the students.

The author has made every attempt to ensure that the information presented is current and that all recommended sources are accurate.

Veterinary medicine is a service profession. It requires flexibility, inspiration, humility, compromise, a sense of humor, and an ironclad work ethic. I hope this text engages and challenges the student to embrace all aspects of a rewarding career.

ONLINE SUPPLEMENTS ACCOMPANYING THE TEXT

An online Instructor's Manual and MyTest are available to instructors at www.pearsonhighered.com. Instructors can search for a test by author, title, ISBN, or by selecting the appropriate discipline from the pull-down menu at the top of catalog home page. To access supplementary materials online, instructors need to request an instructor access code. Go to www.pearsonhighered.com, click the Instructor Resource Center link, and then click Register Today for an instructor access code. Within 48 hours after registering, you will receive a confirming e-mail including an instructor access code. Once you have received your code, go to the site and log on for full instruction on downloading the materials you wish to use.

ACKNOWLEDGMENTS

I wish to thank the following people for their contributions and support in the development of this book.

Tim Ellis, DVM—Mid Rivers Equine Center, Wentzville, MO
Harvest Plaza Animal Hospital, St. Charles MO—Laurie Schmidt; Brittany Pfeifer; Caroline Sandholm, RVT; Carla Paul; Theresa Knoblock
Ray Higley, B.A., M.Ed
Michael Richards, DVM
Thomas Sigel—thank you for your patience and support
Staff of Pearson Education—Bill Lawrensen, Lara Dimmick, Kris Roach, Alicia Wozniak, and Katrina Ostler with Cenveo Publishing Services
St. Louis Veterinary Emergency Group

I would also like to thank the reviewers for their thoughtful comments and suggestions. They are Kathleen M. Corcoran, DVM, Cuyahoga Community College; Laura Earle, DVM, Brevard Community College; Angela Hutchinson, Sanford-Brown College; Leslie Sinn, NOVA Veterinary Technology Program; Melissa Stacy, Rockford Career College; Peg Villanueva, DVM, MPH, Vet Tech Institute – Indianapolis; Jennifer Wells, DVM, University of Cincinnati, Blue Ash College; Catherine Reid, DVM, Laguardia Community College; Angela Beal, Vet Tech Institute at Bradford School; Tricia Stobbs, Globe Education Network; Sarah Bjorstrom, Globe University; Mary M. Hatfield, Lincoln Memorial University; Brian L. Hoefs, DVM, Globe University; Kelly Black, Cedar Valley College; Anne Duffy, Kirkwood Community College; and Mary O'Horo Loomis, DVM, SUNY Canton.

1

The Veterinary Profession and Its Role

Chapter Outline

Learning Objectives

At the end of the chapter, you should be able to:

- Compare and contrast the four guiding principles of veterinary medicine.
- Define the study of veterinary technology.
- Explain the responsibilities of the veterinary technician.
- Give five examples of career paths of graduate veterinary technicians.
- Describe the contributions of the veterinary profession to animal and human health.
- Identify the associations that represent veterinary medicine, and discuss their purpose as it relates to both animal health and the support of veterinary professionals.
- Summarize the licensing process and continuing education requirement for graduates of veterinary technology programs.
- Explain how the principles of professionalism can be incorporated into acceptable work habits, human relations, and communication.

INTRODUCTION

Why Do You Want to Study Veterinary Technology?

Your Personal Goals   When asked "Why do you want to study Veterinary Technology?" most students enthusiastically reply, "because I love animals and want to help them" or "I love animals more than people." Do these statements mirror your own feelings? Before you begin your formal studies, ask yourself why you want to become a veterinary technician. Take a few moments to carefully examine your feelings and write down three reasons why you want to study veterinary technology.

Great! Now you have defined three goals for yourself. Let's compare them to some of the guiding principles of veterinary medicine.

Goals of the Profession

- **Protecting animal health and welfare.** Veterinary professionals devote themselves to the study, diagnosis, treatment, and prevention of animal diseases. They advocate for animal welfare by protecting their physical and psychological well-being.
- **Protecting the public health.** As you will learn, there are many diseases that are transmissible from animals to humans. As a veterinary professional, you also become a guardian of public health by recognizing and implementing measures to contain the spread of disease.
- **Public education.** Teaching people how to properly care for a pet is a core responsibility of the veterinary staff. You will help clients with information on immunizations, nutrition, behavior, parasite control, and many other aspects of animal health. In addition, you will have the opportunity to counsel your clients regarding the selection of pets that are appropriate to their lifestyle and resources. Many species are appealing to own but are inappropriate as pets in inexperienced hands. Averting a poor outcome for the animal and the owner is one goal of your education.
- **Protecting the nation's food supply.** One responsibility of veterinary personnel is to ensure that milk, meat, and eggs are safe for public consumption.
- **Conservation of wild animal populations and their environments.** Loss of habitat, hunting, and poaching endanger many animals. Conservation of land, resources, and legal protection is a role in which veterinary professionals have a voice.

The adorable appearance of animals often motivates you to care for them.
Eric Isselée / Fotolia LLC

Veterinary professionals care for the needs of numerous species.
Anna Kucherova / Fotolia LLC

Veterinary personnel are employed to inspect and safeguard the nation's food supply.
Alaettin YILDIRIM / Shutterstock

The Bengal tiger is endangered due to the loss of its native habitat and poaching.
nialat / Fotolia LLC

How do these foundations of veterinary medicine compare with your owns goals? Do you see opportunity in areas to serve people and animals that you may not have previously considered?

WHAT IS VETERINARY TECHNOLOGY?

Veterinary technology is in-depth professional training that prepares you to assist a licensed Doctor of Veterinary Medicine (veterinarian) in the care and treatment of animal patients. Your job will encompass many disciplines, and in the course of your training you will learn the fundamentals of anatomy, physiology, radiology, pharmacology, nursing, dentistry, lab procedures, anesthesia, and surgery in multiple species. A well-trained veterinary technician is invaluable, often serving as the veterinarian's "right hand," and plays a key role in patient care and client education. The job is demanding, and unexpected situations often arise, giving you varied and unanticipated challenges.

Your field is unique. There is no other career that allows you to move freely from surgery to pharmacy to the lab to the surgical suite, participating in all the activities involved in each area. Human medical disciplines are typically limited to training in only one of the many areas of the hospital. Not only will you transition with ease between these various tasks, but you'll also experience the challenge of providing care for many types of animals. Your day might start out with a dog and end with an iguana. That won't happen in a physician's office!

Exotic species provide opportunities to learn new techniques and treatments.
Julie Keen / Shutterstock

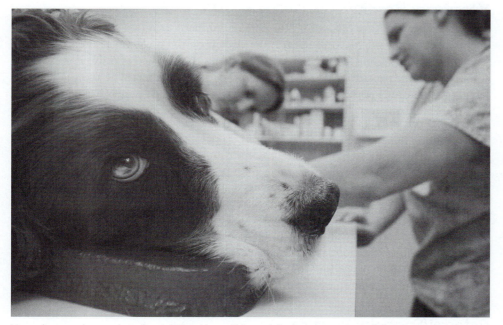

Your day may have a lot of variability. You will have jobs in many areas of the practice and see a variety of patients.
CREATISTA / Shutterstock

WHAT ARE MY RESPONSIBILITIES AS A VETERINARY TECHNICIAN?

Veterinary technicians are allowed to perform most of the vital functions in a veterinary hospital. However, you will not be allowed to diagnose, prescribe medications, perform surgery, or give a prognosis. Additionally, your state may regulate certain tasks.

Technicians follow medical orders, perform patient assessments, give medications, administer treatments, assist in surgery, monitor anesthesia, educate clients, and perform physical therapy and rehabilitation. You will also be required to keep medical records and keep the hospital clean and stocked. Veterinary technicians carry out these duties under different levels of supervision; that is, there are certain permissible tasks, but others that can be performed only under the direct supervision of a licensed veterinarian. You can perform some tasks at any time whether or not the veterinarian is on the premises. The Veterinary Medical Board or equivalent licensing agency in your state regulates these tasks and levels of supervision.

WHAT ARE MY CAREER OPPORTUNITIES?

As a graduate of a veterinary technology program, there will be many opportunities and career paths. You may discover, as many students do, that you find an area of interest in school that you never anticipated. You may develop a love affair with horses, find out you have a head for research, or become fascinated with exotic animal medicine. The good news is that there are avenues to pursue these interests and to expand and continue your education.

Veterinary Hospitals

Veterinary hospitals are focused on the care of companion animals. The scope of species any veterinary practice may treat is dependent on the preferences and skills of the veterinarian. Some practices may limit themselves to one species only. For example, some veterinary practitioners see only feline patients. Others will accept the challenge of any animal patient that comes through the door. As a member of the health care team, you will develop closer client-patient relationships as you see your clients return year after year with their pets for care.

Emergency Clinics

The patients that are presented to an emergency service typically need immediate medical or surgical services. Emergency clinics are fast paced and see a variety of cases without the advantage of having a scheduled appointment. The pace of the workload can change dramatically and is often frantic. Shifts are usually long and include nights, weekends, and holidays. Because clients do not utilize emergency clinics for long-term management or for wellness, you will not build lasting relationships with clients, but you will experience a challenging caseload.

Referral/Specialty Practice

Like many disciplines of human medicine, veterinary medicine is specializing. It is now common to find veterinary practitioners that specialize in only one area of treatment. Dermatology, dentistry, ophthalmology, behavior, surgery, and radiology are just a few of the veterinary specialties. General practitioners often refer patients to a specialist, as they may find it necessary to send a patient for treatments or diagnostics that are beyond the scope of their practice. For example, if a veterinarian is treating a patient that suffers from seizures, they might require an MRI or a CAT scan. It is likely that they will refer the patient to a practice that has this sophisticated imaging equipment. Veterinarians may refer patients requiring long periods of hospitalization or critical care to a specialty practice for monitoring and treatment of specific needs.

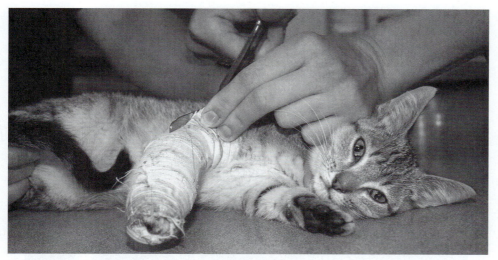

Providing care for sick and injured animals is compassionate and rewarding.
Julie Keen / Shutterstock

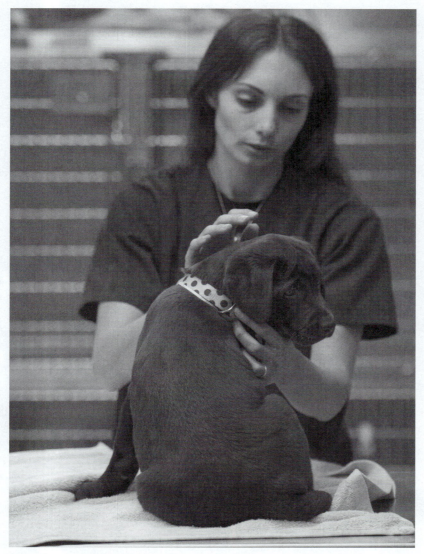

Your job responsibilities will be determined by your state regulatory agency.
aspen rock / Shutterstock

Government

The government employs veterinary technicians in the many agencies that conduct animal disease research, protect the food supply, monitor for emerging diseases, fight bioterrorism and train dogs for search and rescue. There are opportunities both nation- and worldwide.

Military

The Armed Forces employ veterinary technicians to care for working animals, as well as those who may belong to base personnel. There also are civilian positions, so you do not have to enlist for service to work within the framework of the military.

wellphoto / Fotolia LLC

Rescue/Shelter Medicine

Animal rescue organizations and shelters house animals that require behavior evaluation and medical care before they are adopted or retired. The nationwide trend toward no-kill shelters means that the shelters will house more animals, some for a very long period of time. These animals will require regular social interactions and enrichment to meet their physical and psychological needs. Shelters are not just for dogs and cats. Horses, cows, rabbits, goats, chickens, and donkeys are examples of the animals coming from substandard conditions that organizations rescue and rehabilitate.

Wildlife Management and Rehabilitation

Native wildlife often finds conflict in coexisting with humans, and animals can become orphaned or injured by machinery, people, or domestic pets. Wildlife rehabilitation centers focus on restoring these animals to function and releasing them back to their natural habitat. Nonnative species in roadside exhibits, small zoos, and

Shelter animals have physical, psychological, and social needs.
Bine / Shutterstock

Care and rehabilitation of various wildlife species requires special training to meet their individual needs.
Margaret M Stewart / Shutterstock

inappropriate conditions in private homes are often in need of proper housing, behavioral enrichment, and proper nutrition and husbandry. In contrast, nonnative species released into local areas can decimate populations of indigenous wildlife. Developing wildlife management strategies, rehabilitation and humane relocation techniques are needed for controlling these problems.

Zoos/Aquariums

Veterinary staff oversee animals that are in exhibits for public viewing and education. They closely monitor and supervise them for any signs of illness. The veterinary team is also involved with nutrition, husbandry, and breeding programs. Zoos make it possible for people to see animals that they would not have an opportunity to see in their native habitats. These encounters lead to an awareness of welfare and environment, which hopefully results in conservation of species.

Biomedical Research

Medical research uses animal models under the close supervision of veterinary health care teams. They write and enforce protocols for the care and use of animals in facilities that utilize animal models of disease. Students often overlook research as a career path, but it provides the opportunity to work with many species that you might not otherwise encounter and the ability to contribute to research that may help protect both animal and human health.

Zoo animals educate the public about wildlife and conservation.
donvanstaden / Fotolia LLC

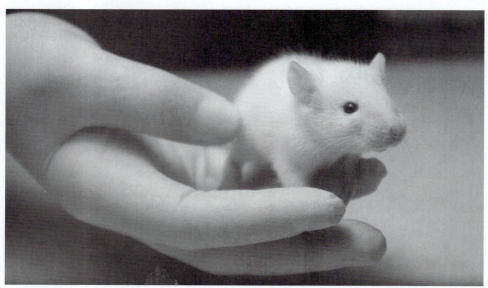

Animals in research provide valuable information about human and animal health.
lculig / Fotolia LLC

Teaching

As you are engaged in the process of learning, your instructors are there to provide you with information and guidance to help you through your veterinary education. The process of teaching is very challenging and rewarding. Once you have completed your education and several years of experience in the workplace, you may find that you too have the desire to teach. Veterinary technology programs employ licensed technicians to provide classroom and laboratory instruction to students.

Diagnostic Laboratories

Many veterinary facilities employ the services of off-site laboratories to process samples for laboratory diagnosis. Busy veterinary hospitals often find it cost effective and a savings of time to send all samples to diagnostic laboratories for analysis. Another benefit is consistency. The same machine analyzes all samples in the same manner every time, which diminishes the number of human errors. Veterinary technicians work in these settings to perform analysis on submitted samples.

Veterinary Product Sales

Veterinary pharmaceutical, feed and supply companies employ veterinary technicians to provide service and sell products related to animal health, nutrition, grooming, and activity. These companies seek individuals who have appropriate backgrounds and knowledge of products that promote animal health.

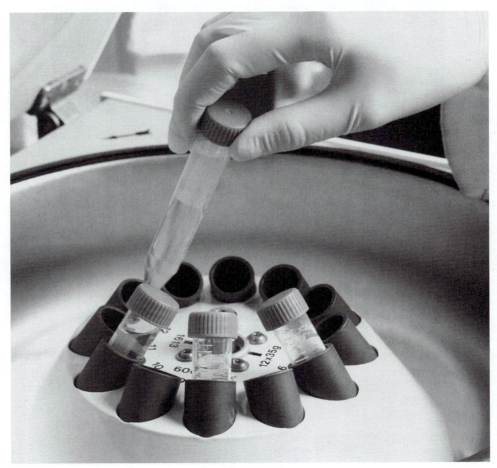

You will learn to use a wide variety of laboratory equipment.
phloxii / Shutterstock

Jim Parkin / Shutterstock

Animal Control

Veterinary technicians can seek employment as animal control officers working alongside law enforcement to address issues of animal welfare and abuse. These individuals are often given police power to enforce the laws and statutes. This job will involve extensive contact with the public. Skill in conflict resolution is a plus in this career path.

Career Websites

While you are beginning your training, take a look at some of the websites that feature job postings for veterinary technicians. You might find just what you're looking for:

USA Jobs website

VSPN organization website

WhereTechsConnect website

American Veterinary Medical Association webstie

National Association of Veterinary Technicians website

IN WHAT FIELDS CAN I SPECIALIZE?

You may find that after graduation you wish to seek advanced training and skill in an area of specialization. The National Association of Veterinary Technicians in America (NAVTA) has developed the Committee on Veterinary Technician Specialties (CVTS). The committee has established areas in which a veterinary technician can specialize. Individuals completing the criteria for advanced skills and training can earn a Veterinary Technician Specialty designation. Academies approved by NAVTA are:

- Dentistry
- Equine
- Anesthesia
- Behavior

- Emergency and Critical Care
- Internal Medicine
- Zoological Medicine
- Surgical Technology
- Nutrition
- Technicians in Clinical Practice
- Clinical Pathology (NAVTA, 2012a).

For more information, visit the NAVTA website.

Reality Check

Before we go any further, we need to address some assumptions that you may have made about your career. It's important that you begin with a solid footing, and it's time to discuss some realities of life as a veterinary professional.

1. **"If I work with animals, I don't have to work with people."** False! You will have teachers, classmates, employers, supervisors, coworkers, clients, and colleagues with whom you will have to maintain cordial, cooperative, and professional relationships. You will need to be able to develop and maintain good working relationships and treat each person with respect and courtesy.

2. **"Animals don't care what I look like."** True! But your clients do. The first part of a trusting relationship is your professional appearance and attitude. How you look, dress, and speak is very important in fostering positive trusting relationships with your coworkers and clients. Your appearance is your first assurance to your clients that you will give their pet your best care. Keep an extra uniform at work. Make sure that your footwear is clean and in good repair, and always practice good hygiene.

3. **"My love of animals is all I need to be a great vet tech."** False! There is a big difference between people who *want* to work with animals and those who *can* work with animals. You must also love science. Your love of animals is the motivation for your education, but you must open yourself to the *science* of how to care for them. You must be able to set aside your emotions and provide assistance and care for patients who by and large do *not* want you to do your job. You must deal with the realization that many of your patients will be anxious, nervous, and aggressive.

4. **"Owners will take care of their animals."** Maybe. For a variety of reasons—emotional, moral, ethical, religious, and financial—an owner may not choose a recommended course of treatment for his or her pet. You may feel that the patient is not receiving optimum care. Some owners may decide on euthanasia. Our job is to educate, not pass judgment.

5. **"I can do everything I used to do plus school."** False. You will need to practice time management. You may need to postpone for a while some of the extra things you did in your spare time. You will need to read and prepare for every class period and laboratory exercise. You will get so much more out of your education by doing your reading and assignments prior to coming to class. This will help you prepare to ask questions and allow your instructor to clarify information. You should estimate two hours of study for every hour of class time. Your education is ultimately your responsibility.

6. **"I was told there would be no math."** False! There *will* be math! You will be responsible for learning how to calculate drugs doses, fluid rates, and mixing solutions, among other calculations. You may find math very intimidating—many students do. The good news is that you are probably doing math everyday without realizing it. If you shop for bargains, use coupons, or plan food for a

You will need to prepare for class by budgeting time to study.
Solphoto / Shutterstock

Math is an essential component of your training.
Roger Ashford / Fotolia LLC

Cleaning is a necessity and a job expectation.
karakotsya / Shutterstock

party, you are using the same type of math skills that you will need to do the calculations your job requires.

7. **"Once I am a finished with school, I am finished with cleaning kennels, mopping floors, and front desk work."** False! Despite your degree and licensing, cleaning the kennels will most certainly be part of your job. Truthfully, cleaning will be a way of life for you in the workplace. Cleanliness and order are necessary for kennels, bathrooms, laundry, break rooms, and exam rooms. Washing and caring for surgical instruments and the maintenance of other equipment is all part of the job. You will also have your fair share of clerical duties. There will be records to file, phone calls to return, and reminders to mail. Be ready to participate in all of the activities that keep a hospital clean, efficient, and focused on patient care.

THE HUMAN-ANIMAL BOND

Imagine yourself watching a litter of playful puppies or kittens. You feel happy and delighted with the way that they frolic and play. Their movements, body postures, and facial expressions are a joy to watch. You want to interact with, touch, and hold them. You have their best interests at heart.

Benjamin Simeneta / Fotolia LLC

People across all cultures invite animals to become part of their world. The animals may serve many purposes, and recorded history has documented the interactions between animals and people. We refer to the mutually beneficial relationship shared between people and animals as the **human-animal bond**.

This bond may have been a significant motivation for you to seek a career as a veterinary professional. The reasons can be limitless. What brought you to this point? Perhaps you had a dog or cat that provided companionship and comfort. Maybe it was a horse that took you on long rides in the country, raced you around a barrel, or jumped over a rail. You felt connected, like you were one being. If you were raised in a rural area, you might have raised cattle, goats, sheep, chickens, or other farm animals. You were concerned for their health and welfare every day for the value they contributed to the family farm or the milk and meat that they provided. Perhaps you are mesmerized by the beauty of animals in their natural habitats.

All of these scenarios describe ways in which animals and people connect in mutually beneficial ways. As you advance in your profession, people will tell you

Early cultures recorded their observations of and interactions with animals by painting their images on cave walls.
Pichugin Dmitry / Shutterstock

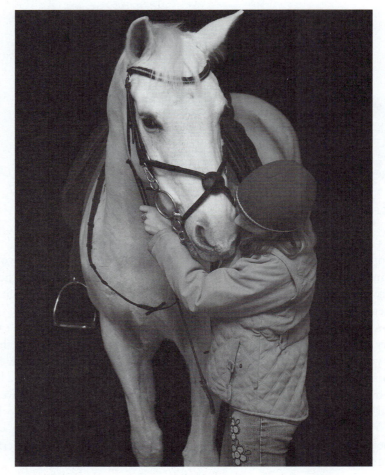

The bond between humans and animals is powerful and emotional.
Andrey Armyagov / Shutterstock

Appreciation of the beauty of wildlife is part of the human-animal bond.
Mircea BEZERGHEANU / Shutterstock

how their lives are enriched and enhanced by the presence of animals. Many will tell you that they consider their pets family members, and often they make great allowances to accommodate the needs of those pets. Pet owners will changes their habits, homes, schedules, and even financial stability to make sure that they feed, exercise, and enrich their animals with social experience. In return, pet owners feel as if the pets love them unconditionally and without judgment. They have constant and comfortable companionship. Pets accompany their owners through some of life's worst traumas and can be a stabilizing influence when everything else seems to go wrong. The feeling that you experience when you are connected, concerned, and completed by the presence of animals is the human-animal bond. It is what has brought you to the decision to devote your life to the care of animals.

HISTORY OF THE VETERINARY PROFESSION

Before we jump into the science of veterinary medicine, we need to spend a little time understanding the roots of our profession. It will help you to understand how veterinary medicine has evolved over the centuries to the career as you experience it today.

In 2011, veterinary medicine celebrated its 250th birthday. The first school of veterinary medicine opened in Lyon, France, in 1761, marking the beginning of veterinary education. As a new profession, the primary concern of veterinarians was the care of horses, mules, and cattle. These animals were the backbone of transportation, agriculture, and food production, so their health was vitally important. The

Vladimir Voronin / Fotolia LLC

People treasure the comfort and companionship of their pets.
Inferna / Fotolia LLC

Nate Allred / Shutterstock

Kletr / Shutterstock

TABLE 1.1
EXAMPLE OF LEVELS OF SUPERVISION: MISSOURI VETERINARY PRACTICE ACT

Immediate Supervision	Direct Supervision	Indirect Supervision
The veterinarian can see and hear what the veterinary technician is doing.	The veterinarian is on the premises and is available if needed.	The veterinarian is not on the premises, but the veterinary technician has direct communication.
Examples of tasks requiring immediate supervision: Anesthetic induction Surgical assistance	Examples of tasks requiring direct supervision: Dental prophylaxis	Examples of tasks requiring indirect supervision: Administering medications and treatments Collection of laboratory samples such as blood and urine Initiating life support after direct communication with the veterinarian

Source: Adapted from Missouri Veterinary Medical Practice on the Missouri Division of Professional Registration website.

care of animals was often a challenging task for veterinarians, because the resources of medicine that we take for granted were not yet available. Antibiotics and anesthetics were still in the future, and surgical and medical treatments were crude and often ineffective because there was not yet a full understanding of animal diseases (Animal Pet Doctor, 2002).

In the late 1800s, there were important advancements in disease diagnosis and vaccine development. By 1892, 14 diseases had been eliminated from livestock, poultry, and horse populations in the United States (King, 2006). See Table 1.1. Veterinarians contributed to the science of human immunizations through work isolating infectious organisms in animals. Testing cattle for tuberculosis had begun, which helped stop the spread of this contagious disease to other animal populations and humans. In response to diseases and epidemics in livestock populations, **The Bureau of Animal Industry** was established in 1884 to protect people from contamination in the food supply and to improve the health of livestock (National Archives, 2012).

During these early years, veterinary medicine was exclusively a male profession, and not a glamorous one. Society considered it a dangerous and dirty profession ("man's work"), and women were not readily accepted into veterinary training. A shift of this dynamic began in the 1970s as new laws made it illegal to deny women access to veterinary school. Additionally, as more drugs became available for safe patient sedation and anesthesia, restraint of large or aggressive species became easier. In 2011, women made up 78 percent of incoming veterinary students and female veterinarians outnumbered their male colleagues in practice (Tremain, 2010).

The practice emphasis of veterinary medicine has changed as well. While the veterinarian's main focus used to be livestock management, the emphasis of today's veterinarian is primarily the health of companion animals. The modern veterinarian's patients are more often family pets and "exotic" animals. Equine and livestock health is still a very important component of the profession, but the majority of veterinarians today practice companion animal medicine (AVMA, 2012).

The majority of veterinarians in training are female.
Goodluz / Shutterstock

Contributions of Veterinary Medicine to Public Health

Veterinary medicine has made numerous contributions to public health. Training in the areas of parasitology, zoonotic diseases, research animals, and agriculture enables veterinary medicine to bridge the gap between human and animal disease. Because people live and work in close proximity with animals, they are often the

gorillaimages / Shutterstock

victims of diseases that originate in animal populations. Studying animal disease models allows both veterinarians and physicians to understand the causes of diseases and how they transmit. This is a significant contribution to human health, because parasites, viruses, and bacteria harbored in animal populations have been responsible for causing some of the worst plagues in history, killing millions of people (NCBI, 2011).

Rat fleas spread bubonic plaque, which killed 25 million people in Europe in the fourteenth century.
Heiko Kiera / Fotolia LLC

It is important as a veterinary professional that you understand that veterinary medicine not only serves the needs of animals, but is also a guardian of public health. That responsibility continues today. Veterinary professionals are involved at many levels to ensure public health and safety. Some of the responsibilities include:

- Protecting and ensuring a safe and healthy food supply
- Monitoring animal populations for contagious/infectious disease
- Surveillance of bioterrorism
- Disease research

The One Health Initiative has recognized the impact of animal health on human health. To date, we have always studied human medicine and veterinary medicine as separate disciplines. There is little collaboration between veterinarians and physicians while in medical school or in practice. Recognizing the important overlap, the One Health Initiative encourages cooperation among all health disciplines to improve public health, biomedical research, and the scientific knowledge base. Explore the One Health Initiative and its integrative approach to uniting human and veterinary medicine at their website (One Health Initiative, 2012).

You are joining a profession that has helped to identify, control, or eradicate rabies, hog cholera, tuberculosis, foot and mouth disease, mad cow disease, salmonella, brucellosis, blackleg, scrapie, influenza, and rinderpest. You may not be familiar with many of these diseases. With the exception of rabies, chances are you've never heard their names.

Hog cholera, or **swine fever**, is a viral infection of pigs, causing fever, diarrhea, and inflammation of the blood vessels. Researchers identified the causative organism and produced a vaccination. Hog cholera was eradicated in the United States in 1978, saving the swine industry over $100 million annually.

Brucellosis is a contagious bacterial infection of several species, including dogs, pigs, and sheep; however, the disease has its greatest economic impact in cattle, causing weight loss, lameness, and late-term abortion. In humans, brucellosis causes muscle and joint pain and severe fever, which may wax and wane for years. The establishment of the Brucellosis Eradication Program is a cooperative effort among individual states and the federal government to identify and cull affected livestock. As of 2000, 44 states are brucellosis-free.

Blackleg is a contagious and fatal disease of young cattle, caused by bacteria (*Clostridium chauvoei*) that live in the soil. Young animals ingest the bacteria, where it gains entrance through small breaks in mucous membranes and the digestive tract. Affected animals first develop lameness, followed by gas swellings under the skin. They die quickly, usually in 12–48 hours.

This is just a small sample of the contributions from dedicated veterinary health professionals. See Table 1.2 for additional listings.

TABLE 1.2
DISEASES CONTROLLED BY VETERINARY RESEARCH

Disease	Cause	Clinical Sign
Rabies	Rhabdovirus	Paralysis, brain inflammation, death
Hog Cholera	Pestivirus	Fever, inappetence, death
Tuberculosis	*Mycobacterium spp.*	Emaciation, wasting, death
Foot and Mouth Disease	Apthovirus	Ulcerated sores on feet and mouth
Mad Cow Disease	Prions	Fatal encephalopathy
Salmonella	*Salmonella sp.*	Diarrhea, dehydration, fever
Brucellosis	*Brucella abortus*	Late-term abortion in cattle
Blackleg	*Clostridium sp*	Gas gangrene
Scrapie	Prions	Fatal encephalopathy
Influenza	Viral	Pneumonia

Controlling disease in livestock is important. Fewer losses to disease means a more affordable and safer food supply for consumers.

WHAT ARE THE GOVERNMENT AGENCIES THAT OVERSEE THE PRACTICE OF VETERINARY MEDICINE?

1. The **United States Department of Agriculture (USDA)**, was founded in 1862 and conducts research into animal welfare and management, veterinary biologics, and controlling animal disease. It has a large regulatory role in the practice of veterinary medicine. The following agencies are divisions of the USDA and are responsible for food safety and the protection of animal health and welfare (USDA, 2013c).

   a. The **Food Safety and Inspection Service (FSIS)** inspects meat, eggs, milk, and poultry to check their safety before they enter the food supply. The FSIS employs veterinarians and veterinary support personnel at food-producing facilities to ensure that safe practices are followed and at slaughterhouses to enforce the Humane Slaughter Method Act (USDA, 2013a).

   b. The **Animal and Plant Health Inspection Service (APHIS)** is responsible for protecting and promoting animal health and for the administration of the **Animal Welfare Act**. Congress passed the Animal Welfare Act in 1966. It sets standards for the care, treatment, housing, nutrition, and provision of other physical and psychological needs of animals used in research, exhibition, or commercial purposes. Facilities that use animals in this manner must license themselves with APHIS. The AWA also prohibits the use of animals in baiting or fighting (USDA, 2013b).

2. The **Environmental Protection Agency (EPA)** develops and enforces regulations regarding the use and disposal of chemicals that can harm people, animals, and the environment. How does this affect veterinary medicine? The use of **pesticides** in companion animals to control fleas and ticks is a familiar example. These chemicals must pass rigid standards of testing before they receive a label that marks them safe for use. We also use pesticides to control pests on cattle that cause irritation and spread disease. The EPA also monitors water quality in areas surrounding factory farms, to detect contamination from waste runoff (EPA, 2013).

3. The **Food and Drug Administration (FDA)** houses a division, the **Center for Veterinary Medicine (CVM)**, that regulates the production and distribution of drugs that we give to food and companion animals. These include, but are not limited to vaccines, food additives, and nutritional supplements. An important component to the regulatory process of the CVM is establishing guidelines for the approval process for the use of antibiotics and other drugs in food animals. A "withdrawal time" is part of this approval process—the number of days an animal must remain out of the food supply until a drug has metabolized and cleared from the system. This prevents the compromise of human health and safety. A stringent monitoring system for drug residues is necessary to ensure compliance (FDA, 2013).

4. The **Drug Enforcement Administration (DEA)** enforces the laws that regulate the use of controlled substances. The agency classifies controlled substances based on their potential for addiction and abuse. As a veterinary technician, you may have access to the controlled drugs used to provide anesthesia and manage pain. You will need to keep extensive and accurate records to document each dose of controlled

substance that the hospital uses. The DEA can request hospital records at any time for an audit (DEA, 2012).

NATIONAL ASSOCIATIONS

The American Veterinary Medical Association

The AVMA is a non-profit association representing veterinarians in many different career disciplines. The objective of the Association is "to advance the science and art of veterinary medicine, including its relationship to public health, biological science, and agriculture."

The AVMA Committee for Veterinary Technician Education and Activities (CVTEA) is "a program of accreditation of training for animal technicians." The objective of the committee is "to recognize veterinary technician training programs that are fully capable of graduating acceptable assistants for veterinarians and to assist in the development of such programs." (AVMA, 2013).

NAVTA: The National Association of Veterinary Technicians in America

The **National Association of Veterinary Technicians in America (NAVTA)** is the organization that represents and promotes the profession of veterinary technology. NAVTA was founded in 1981, with the goal of advancing high standards of care and humane treatment of animals. It sponsors continuing education and veterinary technician specialty training. In recognition of your extensive training and education, NAVTA sponsors National Veterinary Technician Week annually in October. The purpose of this special celebration is not only to recognize your hard work and dedication, but also to raise public awareness about the scope of your job. NAVTA has also drafted the **Code of Ethics** for veterinary technicians. It comprises the guiding principles for the practice of veterinary technology and sets the standard for your professional behavior (NAVTA, 2012b).

You should familiarize yourself with the NAVTA website and the information and support it offers veterinary technicians.

STATE AND LOCAL ASSOCIATIONS

Veterinary Medical Boards

Every state has its own method of regulating the practice of veterinary medicine. Most states have veterinary medical boards, but others accomplish the same purpose with **divisions of professional licensing**. They supervise the licensing of veterinarians and veterinary technicians, issue facility permits, and enforce current laws and statutes. They are also responsible for the resolution of complaints clients lodge against veterinarians or veterinary hospitals.

Veterinary Medical Associations

Along with the veterinary medical board, many states have their own veterinary medical associations. These groups promote animal health and welfare in their states. Members meet on a regular basis to discuss current topics in the veterinary community. Disease recognition and treatment, new diagnostic equipment, and principles of practice management are common topics at these meetings. Many of the state associations have a separate organization for veterinary technicians and students. This is a great way for you to meet colleagues, share information and expertise, and have fun. State associations promote social gatherings and help build a support network of friends and mentors. They frequently sponsor seminars to meet the need for continuing education of veterinarians and veterinary technicians. Involvement with veterinary medicine at the state level often opens doors for participation at the national level. If you're interested in the regulatory, legislative, or political side of the profession, membership in the state association is a good starting point.

At the local level, the city or county in which you live may have its own veterinary association. Local associations provide networking opportunities for veterinary professionals in a close geographical area. Monthly meetings, newsletters, speakers, and demonstrations provide a means of keeping veterinary professionals up to date and also give people a chance to meet on a regular basis to share social time and camaraderie. There are many benefits to joining a local veterinary association, and all members of the veterinary team are welcome.

LICENSING PROCESS AND CONTINUING EDUCATION

To obtain your veterinary credentials, you must first satisfy the eligibility requirements of your state. For most states, that means you must successfully complete an AVMA-accredited veterinary technology program. Some states consider alternate educational paths and on-the-job training in determining eligibility.

Throughout your education, you will be exposed to information and skills that will reflect the content of the **Veterinary Technician National Exam (VTNE)**. The VTNE is a comprehensive test that covers a wide range of veterinary concepts and skills. You are eligible to take the exam after you have successfully met the requirements of your state or province. This examination is challenging and requires extensive preparation and review. The **American Association of Veterinary State Boards** owns and administers the exam. You may take the exam once the ASVSB verifies your eligibility. The organization offers the test three times a year (AAVSB, 2012).

Upon successful completion of the VTNE or other equivalencies, you may apply for licensing in any state where you wish to work. Most states require that you pass a state exam that focuses on the laws and practice of veterinary medicine in that state. Each state administers its own exam, and you can be licensed in multiple states. Licensing credentials vary. Some states confer a license (LVT or LVMT), others a registration (RVT) or a certificate (CVT). Only people who hold these credentials may refer to themselves as a "veterinary technician." Prior to your admission to school, your externship placement, or your application to take a state exam, a criminal background check may be required. You may have to explain any criminal offenses or provide court documents. The licensing agency will have the final decision about granting credentials.

Most veterinary technicians require **continuing education credits** to maintain licensure. You receive these credits when you participate in approved activities that keep you up to date on new trends and methods, equipment, and concepts in the field. The number of required credits varies depending on where you practice. This is where your local and state associations are very helpful. You can obtain continuing education in a variety of ways, including:

- Seminars sponsored by pharmaceutical companies
- Online training from veterinary websites
- Journal articles
- Local, state, regional, and national meetings

Because you can obtain continuing education credits from a variety of sources, your state licensing agency may have restrictions on what is and is not acceptable. You should check to make sure that your credits satisfy the continuing education requirement for licensure. The AAVSB issues RACE approval on the content of many seminars and conferences. The purpose of **RACE (Registry of Approved Continuing Education)** approval is to apply a uniform standard to continuing education. This ensures the quality of the continuing education speakers, information, and material that an organization presents.

WHAT DOES IT MEAN TO BE "PROFESSIONAL"?

Principles of Professionalism

Now that you have an idea about your career in veterinary medicine, let's spend a little time on what it means to be "professional." By definition, a professional has received specialized education and training in a field of study. However, professional behavior in the workplace is also an important aspect of your job. You must begin to polish your professional attitude and actions.

Think about your last visit to a doctor's office or a business. You needed someone to listen and address your needs. You wanted it done completely and in a timely manner. You wanted to be treated with courtesy and in a respectful manner. When we talk about what it means to be a professional, you should keep these experiences in mind and use them as a reference point. Let's address some of the important aspects of developing your own professionalism as a veterinary technician.

1. **Appearance.** You must make an effort every day to appear professional and competent. The first step in accomplishing this goal is to take care of your appearance. It is critically important that you dress in an appropriate manner, following your employer's dress code. Your clothing must be in good repair and fit properly. Have enough work clothing for your work week. Coming to work inappropriately dressed because your "scrubs are dirty" is a poor excuse and substandard for the image of your workplace. Wear appropriate footwear, and make sure it is clean. Practice good hygiene, avoiding the use of heavy fragrances. You will most likely be working in close proximity to your coworkers and clients, so be aware that body odors, fragrances, and poor grooming are magnified in this type of situation. Your clients must have a basis upon which to establish a trust with your veterinary practice, and your personal appearance has a big impact on the development of that trust. Remember, the animals don't care what you look like, but your clients do.

2. **Work ethic.** You are about to enter a profession that can't tell time. Veterinary offices are busy places, and the care of animals does not take place according to what time of day it is. All manner of events during a workday can provide challenges, setbacks, and a need to rearrange the day's schedule. A good work ethic means that you will come to work every day that you are scheduled, on time and prepared to do your job the moment you enter the workplace.

A neat, clean appearance tells your client that you are ready for work.
Tyler Olson / Shutterstock

You need to leave all events from your personal life outside the door. Your job is to focus on your patients', clients', and coworkers' needs. Don't bring drama to the clinic. It is bad form to air your personal problems, and it diminishes the quality of care that you provide for your patients.

Flexibility is a must. Emergencies and unexpected events may require that you stay beyond your scheduled shift. Look for work. There are an endless number of mundane tasks that you must perform daily to support veterinary patient care. Laundry, kennel cleaning, autoclaving, record filing, and inventory are just a few examples of the tasks that often need attention between cases or when the work day slows. These are support staff responsibilities.

3. Speech and grammar. The way that you address clients, express your ideas, and convey them in spoken and written form is part of your professional training. The medical terminology that you will learn as you go through school is a tool to help you communicate in a professional manner without excessive emotion. Practice using your medical words in sentences so that they become natural and familiar. Your grammar will need polishing. Avoid common mistakes when speaking. Among the most misspoken phrases are "I seen . . ." and "we was . . ." This sends a message to the client that your education is substandard.

4. Ethical conscience. You will be bound by an oath to the care of your patients when you leave school. There will be many opportunities for you to breach that oath in order to save time or effort. "Ethical conscience" implies that you will uphold your obligations to your patients, clients, coworkers, and employers by performing your job to expected standards, whether someone is observing you or not. This means that you will honestly and faithfully perform all tasks expected of you in an appropriate manner. You will admit and report any errors or mistakes promptly. You will communicate professionally and respectfully with coworkers. You will adhere to the principles of medical confidentiality and will not discuss clients or

Piotr Marcinski / Shutterstock

patients outside of the workplace. You will also not discuss with others treatments or procedures done in your workplace.

5. Sympathy and support. You will encounter many situations in which a patient has a poor prognosis or life-threatening illness. Clients will have to make very difficult choices about their pet care. No matter what path a client may take, it is the staff's responsibility to show support and compassion for all patients and clients during these times. You may find that it is in direct conflict with your own views on how to handle a situation. This is one of the challenges that you will face as a member of the veterinary health care team. You must put aside your personal feelings and offer a few words of support or comfort.

It takes years to develop a good professional demeanor. Begin by practicing some of the characteristics that we just discussed. Fair or not, people form an impression and judge you on how you look, how you communicate (both in writing and speech), and how you convey your respect to others.

Whether you are attending an "on ground" school or you are a distance learner, here are some important tips to help you be a successful veterinary technology student.

1. Look, listen, and feel! This is the time to look at, listen to, and feel every animal you can so that you become familiar with what is normal. It will take a long time before you begin to recognize abnormalities in the way a heart sounds or an abdomen feels. Start now!

2. Form study groups. People learn in different ways. Some students are visual learners; they have to see how to do something before they can learn to do it. Others are auditory learners; they have to hear something before they can understand it. When you study in groups, each student brings his/her own learning style to share. You may have seen something that your classmate missed, and she may have heard something that you didn't.

3. Ask for help the minute you feel confused. Don't wait until several class periods have passed to ask questions or request clarifications.

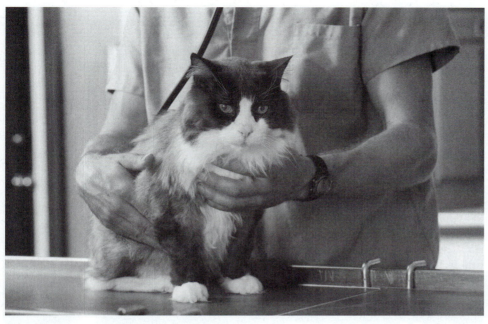

Take the time to practice your skills every day.
Tyler Olson / Shutterstock

4.   Be a mentor. As you become more proficient, offer help and support to your classmates. If you can teach something to someone, it helps you too—you will remember concepts and skills with ease.

5.   Study every day. Don't wait until the last minute to write a paper, turn in a project, or study for an exam.

A study group motivates you and helps you learn new concepts.
Corepics VOF / Shutterstock

Support and teach others as you gain confidence.
Vaju Ariel / Shutterstock

6. **Use online veterinary resources.**

Veterinary Information Network website

VetLearn website

American Veterinary Medical Association website

National Association of Veterinary Technicians in America website

American Animal Hospital Association website

SUMMARY

The veterinary profession has a long and proud history protecting and improving the health of animals and the public. A profession that has its roots in agriculture has grown to incorporate a vision of global health. Your entry into the veterinary profession will provide many new challenges.

Discovering the foundations of your career starts with a basic understanding of the laws and regulations that govern your profession. You must become aware that as a veterinary team member you will be responsible for following state and federal laws.

There are many professional associations and regulatory agencies to provide you with proper guidelines. Use these resources to your benefit: keep current with issues in veterinary medicine and changes that can affect the way you care for your patients. Join professional associations and reap the benefits of connecting to people that love and work with animals. You will take better care of your patients when you can exchange ideas with colleagues.

Your professional behavior is crucial to a successful career. You may know your facts and data, but there is never a substitute for the respect that you extend to your patients, clients, coworkers, and community. That respect is demonstrated by your appearance, your speech and grammar, your work ethic, and your ethical conscience. Just as you model your behavior after your teachers and mentors, future veterinary technicians will model you.

TEST YOUR KNOWLEDGE

1. List and explain five guiding principles of the practice of veterinary medicine.
2. Compare the role of a veterinary technician to that of a nurse working in a human hospital.
3. There are numerous career opportunities in the field of veterinary technology. Name the career field that would include the following duties:
   a. Processing samples to aid in diagnosis
   b. Educating other veterinary professionals about products that support animal health
   c. Restoring animals to health and releasing them to their natural environment
   d. Writing and enforcing protocols in facilities that utilize animal models of disease
   e. Working alongside law enforcement to address issues of animal welfare and abuse
4. What organization is the national representative of the veterinary technician? What are the benefits of membership?
5. Write a paragraph explaining the human-animal bond. Include a personal example of how this bond influenced your decision to pursue veterinary technology as a career.
6. Compare and contrast the primary emphasis of veterinary practice from the 1700s to today.
7. Prepare for your career by discovering what duties veterinary technicians can perform in your state of residence. If applicable, give examples of *levels of supervision.*
8. What organization administers the VTNE and RACE approval of continuing education?
9. Define the term *work ethic.* What characteristics are positive examples of a good work ethic?
10. On Sunday evening, you are caring for hospitalized patients. You accidentally feed a patient that was not supposed to have food or water in preparation for surgery Monday morning. You are in the hospital by yourself and no one witnessed the error. What is the professional standard for managing this error?

BIBLIOGRAPHY

AAVSB. 2012. "About Us." Accessed February 16, 2012 from the American Association of Veterinary State Boards website.

Animal Pet Doctor. 2002. "The History of Veterinary Medicine" Roger Ross, DVM. Accessed May 19, 2012 from The Animal Pet Doctor website.

AVMA. 2012. "Market Research Statistics." Accessed June 1, 2012 from the American Veterinary Medical Association website.

AVMA. 2013. "Who We Are." Accessed June 20, 2013 from the American Veterinary Medical Association website.

DEA. 2012. "DEA Mission Statement." Accessed April 29, 2012 from the United State Drug Enforcement Administration website.

EPA. 2013. "Laws and Regulations." Accessed April 29, 2012 from the Environmental Protection Agency website.

FDA. 2013. "Animal and Veterinary." Accessed May 18, 2012 from the United States Food and Drug Administration website.

King, Lonnie J. 2006. "Veterinary Medicine and Public Health at CDC." *Morbidity and Mortality Weekly Report.* Accessed May 1, 2012

Missouri State Veterinary Medical Board. 2012. "Required Levels of Supervision." Accessed June 4, 2012 from the Missouri Division of Professional Registration website.

NCBI (National Center for Biotechnology Information). 2011. "Plague" Accessed May 30, 2012 from the U.S. National Library of Medicine website.

NAVTA. 2012a. "Committee on Veterinary Technician Specialties." Accessed May 18, 2012 from the National Association of Veterinary Technicians in America website.

NAVTA. 2012b. "Veterinary Technician Code of Ethics." Accessed February 14, 2012 from the National Association of Veterinary Technicians in America website.

One Health Initiative. 2012. "Home Page." Accessed April 30, 2012 from the One Health Initiative website.

Tremain, Jessica. 2010. "Women in Veterinary Medicine". Accessed May 18, 2012 from Veterinary Practice News website.

The U.S. National Archives and Records Administration 2012. "Administrative History." Accessed May 14, 2012. Bureau of Animal Industry Records."

USDA. 2013a. "About FSIS." Accessed April 30, 2012 from the United States Department of Agriculture Food Safety and Inspection Service website.

USDA. 2013b. "About APHIS." Accessed April 29, 2012 from the United States Department of Agriculture Animal and Plant Health Inspection Service website.

USDA. 2013c. "Animal Health" Accessed April 30, 2012 from the United States Department of Agriculture website.

2

Office and Hospital Management

Chapter Outline

Learning Objectives

At the end of the chapter, you should be able to:

- Explain how a mission statement helps define an organization's goals.
- List the steps required for opening a new veterinary practice.
- Compare and contrast different types of veterinary facilities.
- Discuss the functions and activities that occur in the different areas of the veterinary hospital.
- Describe the duties of the different members of the veterinary health team.
- List and explain the purpose of the hospital training manual.
- Explain how advertising and marketing build a client base for the veterinary practice.
- Recognize the importance of communication and customer service as it relates to client satisfaction.

INTRODUCTION

As a student of veterinary technology, you spend much of your time and energy studying the *science* of caring for animals. Your studies are difficult and complex because you are responsible for mastering information about multiple species and disciplines. There is no other health profession that covers the scope of duties of veterinary medicine.

In order to provide care in a setting that offers medical and surgical services to a variety of species, careful planning and practice management are key to running a successful veterinary hospital. This chapter will focus on hospital organization and the number of required tasks to make a hospital efficient, effective, and profitable.

"Thank you for calling All Pets Hospital. This is (your name). How may I help you?" With that simple statement, you have initiated contact and invited the client to come to the clinic for pet care provided by trained staff. You know that veterinary hospitals provide medical and surgical care for pets, but do you know how a veterinary hospital is organized from the ground up? What careful planning, policies, and procedures are in place to make it possible for you to come to work every day and care for animals?

The *business* of the hospital is every bit as important as the practice of medicine, for without a stable, organized management system, even the best veterinarian will face challenges that may hinder the success of the practice.

To help you understand the complexity of starting and running a veterinary practice, let's imagine you have just been introduced to a recent veterinary graduate,

The new sign for All Pets Animal Hospital
glopphy / Fotolia LLC

Dr. Newton. Dr. Newton has dreamed of becoming a veterinarian his entire life. He's just out of school and wants to open his own practice. From beginning to end, let's examine all the steps necessary for Dr. Newton to plan and prepare the practice so he can open the doors for patients at the new All Pets Animal Hospital.

THE MISSION STATEMENT

Every business, organization, or club forms for a reason. It may be to provide a service for profit, organize volunteers, or simply be a group of individuals that share a common interest. A **mission statement** is a brief description of the organization's purpose and what it intends to accomplish. Organizations or businesses use this to determine if they are meeting their goals, and it also provides direction for growth and change. Dr. Newton's mission statement should be simple and should contain goals that are motivational but realistic. It should be a reflection of his values and ethics, as well as those of the members who join his practice.

First, Dr. Newton needs to decide what type of practice he wishes to open and what species of veterinary patients he's going to see. He considers the following mission statements:

"The mission of All Pets Animal Hospital is to provide dedicated, compassionate care to pets by providing high-quality service in a state-of-the-art hospital."

"The mission of All Pets Animal Hospital is to provide medical and surgical services for pets, by a devoted staff, in a loving environment."

"The mission of All Pets Animal Hospital is to provide low-cost spay and neuter services for the four-legged members of our community."

marekuliasz / Shutterstock

You can see how each mission statement defines its objectives. They all indicate the desire to provide medical care for animals, but the first statement indicates an obligation to a state-of-the-art facility. This means that Dr. Newton has to provide the most current equipment, therapies, and treatments available for patients. The hospital will have to devote a significant portion of its space and financial resources to equipment, drugs, and treatment options. The second statement emphasizes the emotional component of the veterinary team. The focus of the practice will be on the staff's caring nature and the manner in which they treat patients and clients. Their goal will not only be excellent patient care, but also impeccable customer service with many little ways to say, "we care." The third includes a component of cost saving for the community's residents and restricts its services to spaying and neutering. The clients coming to this clinic will understand that services are limited and that they may require a referral to other hospitals if the pet's needs extend beyond spaying and neutering. The hospital will only need to be equipped with basic surgical equipment, and the focus of the practice will be on the efficient performance of multiple surgeries every day.

After much deliberation, Dr. Newton decides that statement 2 best reflects his personal values and the way he wants to practice veterinary medicine. Now that the Doctor has stated the mission of the practice, the *hard* work can begin.

We will see that Dr. Newton has his work cut out for him. The first task that he must accomplish is to find a property or space in which to put his new practice. The location of a practice has much to do with the type of practice the veterinarian chooses to run. As Dr. Newton is planning a practice limited to dogs, cats, and exotics, a small freestanding building or even a space in a retail strip mall may provide all the necessary space. Dr. Newton plans on answering his own after-hours and emergency calls and might even make a few house calls. At this time, he does not plan on treating any large animals. That would call for space and extra equipment that he does not want to buy right now.

TYPES OF VETERINARY FACILITIES

Before we go any further, let's examine the different types of veterinary facilities. There are many different types, and each will have special requirements for space and equipment.

Veterinary Hospitals

Veterinary hospitals are staffed and equipped to care for a patient that has a need for wellness, long-term, or emergency care. A veterinary hospital is prepared to provide overnight care and support to nurse a patient through an illness. Hospitals are staffed and equipped to provide most of the basic diagnostic equipment. Laboratory, surgical, treatment, and radiology are services standard to most hospitals.

Veterinary Clinic

A veterinary clinic provides wellness and short-term care for non–life threatening conditions. Clinics do not keep patients that require medical or nursing care overnight. A clinic may be similarly equipped as a hospital in order to provide comprehensive care. In many areas, people use the terms *clinic* and *hospital* interchangeably, with no difference in patient or client service.

Emergency Hospital

A veterinary emergency hospital provides urgent and critical care for sick and injured animals. They do not provide routine health care and are not a substitute for a regular veterinarian's care. Most work in partnership with regular practices to provide continuity of care when unexpected illness or injury happens.

These facilities may be open 24/7, or limited to the hours when regular veterinary clinics are closed. They must be prepared to treat life-threatening emergencies,

This freestanding emergency clinic sees patients only on evenings and weekends when regular veterinary practices are closed.
Amy Wolff

perform immediate surgery, and provide critical care. The regular veterinarian often refers patients so he or she does not have to answer calls after the practice has closed for the day. The doctor on duty at the emergency facility will communicate with the practice owner to let him or her know that one of his or her patients has been treated or hospitalized. After discussing the case, the two doctors will decide if they should keep the patient at the ER, discharge it to the owner, or transfer it back to the practice for further observation and treatment.

Referral Center

A veterinary referral center is a *specialty* practice that diagnoses, hospitalizes, and treats veterinary patients with needs beyond the scope of regular practice. For example, if a veterinarian determines that the patient needs an MRI, ultrasound, cancer treatment, or specialized surgery, he or she might send the animal to a referral practice. Referral centers have equipment, facilities, permits, and drugs that are cost prohibitive for the average practice. Some patients require diagnostics and other treatments that are time consuming, and which the practice cannot accomplish within the time frame of a practice day, so the veterinarian may decide to *refer* the patient to one of these centers.

Mobile Unit

A mobile unit is a self-contained veterinary hospital on wheels. Instead of clients bringing their pets to the hospital, the veterinarian brings the hospital to the client. This is especially helpful for clients who cannot travel or have pets with behavior issues such as fear aggression or conditions such as motion sickness that makes travel to the veterinary facility a difficult prospect. Mobile practices can be very well equipped and include X-ray and surgical equipment.

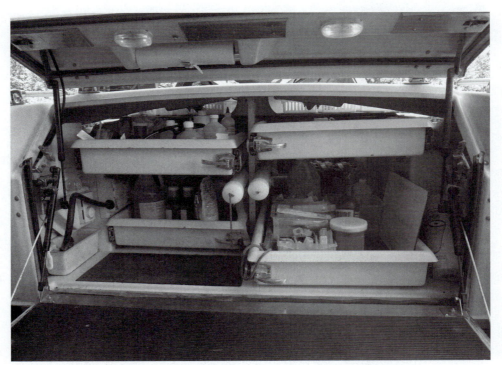

This mobile equine unit is equipped for fieldwork.
Amy Wolff

Small Animal Practice—Exclusive

Small animal practice is limited to companion animals, most commonly dogs and cats. The small animal practitioner may also see rabbits and ferrets and other small mammals, birds, and reptiles.

Large Animal Practice—Exclusive

Large animal practice is limited to cows, horses, and small ruminants such as sheep, goats, and llamas. Large animal practices can have stationary and mobile components. The vet clinic must have enough space for holding cattle or horses in pens or fenced areas. Restraint equipment, such as stocks or chutes, is necessary to safely work with large animals. Stalls and pasture are a must for equine patients. Farm calls are an important part of practice revenue, so trucks equipped with medical and surgical supplies, drugs, and restraint devices are essential. The veterinary team must invest in equipment that is mobile and that they can easily transport. An example is a mobile radiograph unit that the veterinarian can take into a barn or stall to X-ray equine patients for lameness or an ultrasound for pregnancy checks.

Mixed Animal Practice

Mixed practices see both large and small animal patients. This means that the practice must be equipped and staffed to accommodate the needs of both patient types.

Exotic Animal Practice—Exclusive

Exotic practice veterinarians specialize in pets that are "nontraditional," and these come in all shapes and sizes. Each day can hold a different surprise and might include reptiles, primates, birds, small mammals, and large felines. Exotic practices must have drugs and equipment suitable for a variety of species. You may be working with animals for which there is limited knowledge about their diseases,

Amy Wolff

© cynoclub/Fotolia

nutritional needs, husbandry, or behavior. The practice may need to consult with veterinary schools, zoos, and aquariums in order to obtain the best information for treating the patient.

Feline Practice—Exclusive

Feline practices are limited to cats. Cat owners often prefer this arrangement because this way their pets do not encounter dogs or other potentially aggressive animals in the reception area or treatment rooms. Many consider themselves "cat people" and like the idea of a cats-only environment. The clinic space is designed to accommodate feline patient needs. Another advantage is that cats rarely approach a size larger than 20 pounds. This makes it easy to anticipate the patient's needs based on size, and also limits the amount of weight that you have to lift.

Equine Practice—Exclusive

Equine practices are limited to horses. These practices must make a big investment in space and equipment. Stalls and turnout areas will be necessary for equine patient care, and you will need space to observe the animals while they are in motion, to check for the signs of lameness. Proper restraint includes stanchions and padded floors. Horses require special accommodations when undergoing surgical procedures, such as a padded room for inducing anesthesia, a hoist to pick up and transport the patient, and padded surgical tables to protect the animal while you transport it and it is recumbent (lying down).

Zoos and Aquariums

Veterinarians who want to work with wildlife or animals in zoo exhibitions and aquariums must go for training beyond veterinary school. The veterinary hospital is located on the premises with the animals. Zoos and aquariums are the ultimate "exotic" animal practice. Often you and your team will be on the cutting edge of discovering treatments and "best practices" for different species. Observing the physical and psychological signs of health in these animals is critical. There will be challenges and risks in working with wild species, which you might have to sedate and/or restrain in order to provide care. An exciting possibility may be the opportunity for travel to learn about conservation and habitat preservation.

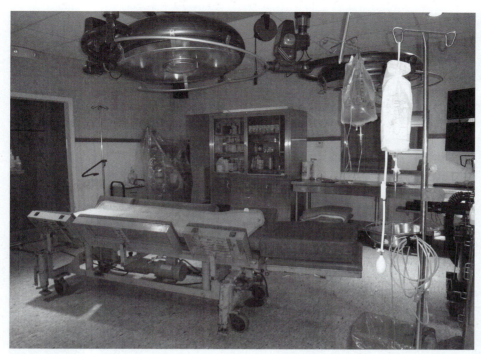

An equine surgical facility will require equipment and space suitable for handling these large animals.
Amy Wolff

The process of opening the doors to All Pets Animal Hospital continues. Now that you have a better understanding of the different types of veterinary practices, we can look at some of the challenges Dr. Newton is facing. He's been busy looking at places to put the practice.

Large stall and turnouts are required for the proper care of equine patients.
Amy Wolff

Amy Wolff

Zoo and aquarium animals require specialized veterinary care.
crlocklear / Fotolia LLC

LOCATION, ZONING, FACILITIES PERMITS, AND FINANCING

Dr. Newton finds a small spot in a strip mall with reasonable rent and a space outside to provide room to walk patients. It's a good location with a storefront that faces the road and is highly visible. Veterinary practices are like any other business. Location is important. Dr. Newton wants the clinic to be highly visible and accessible, have ample parking, and be close to residential areas. The property management company maintains the storefronts and areas around the rental space, which is an added attraction.

Now Dr. Newton has to make sure that he's allowed to put a veterinary practice in the space that he's found. It's important to check with city and county agencies to ensure that the *zoning laws* allow for a veterinary practice. Local governments set zoning laws to regulate the use of land and space. Stop for a moment and think about it: Would you want a busy, 24-hour drive-through restaurant next to your home? Local zoning laws can restrict businesses in residential areas to prevent noise, traffic, bright lights, and other nuisances, preserving the integrity of the neighborhood. The same is true for veterinary hospitals. If Dr. Newton wants to put a veterinary hospital right next to a quiet neighborhood, local residents may object because they don't want to hear dogs barking or may be concerned about sanitation and odors.

Dr. Newton has confirmed that his location is zoned for veterinary practice. His next step is to make inquiries to see if he needs a *business license.* A business license is a permit granted by local government to run a business in a specific location. Not all communities conform to the same laws, so Dr. Newton must do a little research. He visits the website for the Small Business Administration for information to help him determine his needs.

To be eligible to practice, Dr. Newton was required to complete four years of veterinary school and successfully pass the North American Veterinary Licensing Exam (NAVLE) (NBVME, 2012). Dr. Newton lives in a state where a veterinary medical board regulates the practice of veterinary medicine. His state requires that he have the basic equipment and supplies in his hospital to meet the expected standard of care established by the Veterinary Medical Practice Act. This type of facilities permit is required where Dr. Newton wants to practice, but it varies by location. Any new practice would need to check with the state veterinary medical board or licensing agency to see what they require.

Dr. Newton also needs a license from the Drug Enforcement Agency to dispense or prescribe controlled substances. He must renew this license every three years (DEA, 2012).

Financing

Now, it's off to the bank. Dr. Newton will have to apply for a small business loan in order to secure the money he will need to set up the practice. He will need to present a business plan to the loan officer, in the form of a written statement describing his business. This statement describes how he will organize and manage his business and how he will market and advertise. He will need to provide information regarding the size of the community and whether or not there is a large-enough population to sustain a veterinary practice. Dr. Newton will want to research how many people in his area own pets and how much they might spend on pet care every year. He'll also need to offer details on how much money he intends to borrow and how he intends to spend it. He must also discuss how much money he projects he can make at his practice. Doing all of this business homework is

Sven Weber / Fotolia LLC

called *due diligence*, and it demonstrates to the people who will lend Dr. Newton his money that he has properly prepared himself for starting a hospital. The lender will look at all these factors and determine if Dr. Newton and his plans for the new clinic are a good risk for a loan. Dr. Newton makes a list of all the things he can think of that he needs to purchase. The following is a breakdown of expense categories for the new hospital.

The new location of All Pets Animal Hospital.
L Barnwell / Shutterstock

Filling out this application is the first of many steps needed to open the new hospital.
alexskopje / Shutterstock

Veterinarians complete a 4 year program and applicable licensing exams before they can practice.
Andres Rodriguez / Fotolia LLC

The bank will want to see a business plan before they grant the loan for the hospital.
Feng Yu / Shutterstock

Equipment: medical, surgical, laboratory, radiology, dental and treatment areas

Supplies: laboratory, exam room, surgery, dental, radiology, office, janitorial, miscellaneous

Kennels and other containment areas

Anesthesia and monitoring equipment

Drugs

Marketing and advertising

Staff salaries

Office supplies

Office furniture

Prescription pet food

Computers/printers

Books/periodicals and Internet subscriptions to veterinary websites

As Dr. Newton makes his list, he gets a little nervous. It is a big job to organize the list of necessary equipment and supplies to start a new practice. In order to put together the list, he starts by defining the hospital areas. Then, one area at a time, he can determine what equipment and supplies he may need. So Dr. Newton makes a list of the different hospital areas. This is a surprise—he's never really looked at a hospital by its component parts.

Properly sized kennels and containment areas are a must.
Amy Wolff

HOSPITAL AREAS

Reception

This area makes the "first impression" with the client. It should be clean, neat, organized, and free from any odors. The reception area should be both client and pet friendly, with comfortable, spacious seating. Plants, pictures, and bulletin boards warm up the reception area. Some hospitals display pictures of their patients or showcase animals for adoption. TV, magazines, or other entertainment may be provided to occupy clients if a short wait is necessary. This is also a good place to display brochures and other publications that address pet health and wellness. Many important client transactions take place in reception. Clients check in, make payments, and pick up their pets, as well as medications or other care items, so the reception area has to set the *tone* for the clinic. If reception is unkempt, looks disorganized, and has a surly staff at the front desk, the hospital has set itself up for failure. The team members who staff the reception area must be friendly and dedicated to providing flawless

This reception area is neat, clean and welcoming.
Amy Wolff

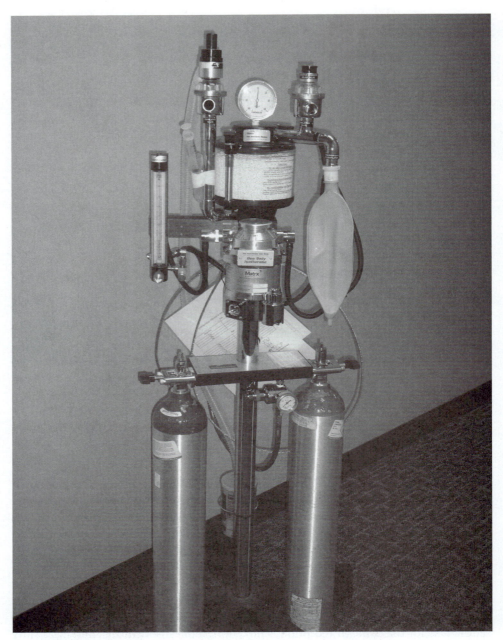

An anesthesia machine.
Amy Wolff

customer service in person and on the phone. The ability to organize, multitask, and even recognize medical conditions that are deemed emergencies are examples of the skills required for taking care of the "front desk."

Treatment

The treatment area of the hospital is where patients are taken for medical or short surgical procedures not requiring an operating room. Treatment areas usually contain sinks or tubs with elevated grates, exam tables, and most of the common hospital equipment and supplies (within easy reach). Examples of treatment area tasks include placing an IV catheter, carrying out a dental prophylaxis (cleaning), administering subcutaneous fluids, cleaning ears, trimming nails, and applying bandages.

Treatment areas should be readily accessible from exam rooms as well as adequately lit and spacious enough to accommodate patients, staff, and support equipment. Cramped spaces can lead to clutter and unsafe working conditions, so treatment areas should be clean and organized.

Hospital Ward

The hospital ward is the area where patients are housed once admitted for treatment or medical or surgical procedures. These areas are equipped to meet the needs of patients that may spend a day or a week in the hospital. A patient may require a fluid pump, heating or cooling devices, frequent nursing care, or physical rehabilitation. The hospital ward may be separate from treatment areas, as animals requiring hospitalization may need a quieter environment, away from the activities, bright lights, and noise of a busy treatment room. Despite the hectic workday, it is important for qualified staff to observe hospitalized patients. A webcam facilitates observing these patients at all times.

Surgery Room/Sterile Prep

The surgery suite (operating room, "OR") and the sterile prep area are often located right next to each other. These rooms are dedicated to surgical procedures, so it is important to take steps to ensure that the staff maintains these areas in a manner that minimizes contamination. The surgery suite usually has only one access door and is located in an area of the hospital that has minimal traffic. The room may require a separate, special ventilation system to provide positive air pressure in the room to keep bacteria and other pathogens out of the surgical environment. Equipment and supplies are "dedicated," which means that you should use them only in the OR. The surgery room is their permanent home, and you should

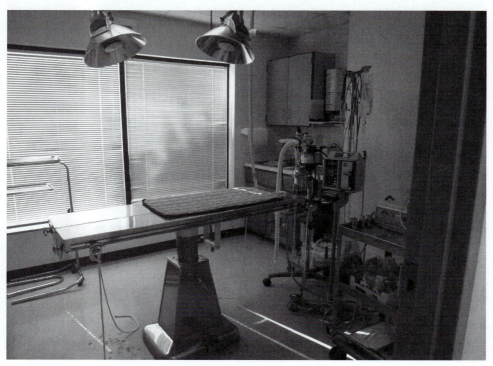

Amy Wolff

not take them into other areas of the hospital. The practice must maintain a stringent cleaning schedule for these rooms, with logs to document the tasks of cleaning and maintenance. There may be restrictions on wearing "street clothes" in the OR. These areas are often "scrubs only," which can include the need for surgical masks and caps.

The **prep area** contains the autoclave and other necessary items to provide equipment and supplies that are sterile and appropriate for use in the surgical patient. It also includes the sink where the surgeon and the assistants will scrub their hands prior to entering the surgical suite. Designated staff will clean and repackage instruments in the prep room. An ultrasonic instrument cleaner removes nonvisible contaminants, thus thoroughly cleaning the instruments and preparing them to go into the autoclave, which then sterilizes them. Cabinets are often "pass through"—that is, open on both sides. You can place sterile supplies in the cabinets from the treatment room and then access them from the same cabinet in the surgical area. There may also be a small, dedicated washer and dryer in the prep room so that you can do surgical laundry separately from clinic laundry.

Radiology/Diagnostic Imaging

Radiology contains the X-ray machine and other equipment that the practice utilizes to produce diagnostic imaging. If the practice uses a conventional system, the X-ray table, radiographic film, and a film processor are located here, as well as personal protective equipment necessary to protect you from exposure to X-rays. Digital X-ray equipment eliminates the need for film. If the X-ray unit utilizes digital imaging, a computer station will be in the room so that the veterinarian or qualified staff can review the images. Depending on the scope of the practice, an ultrasound unit,

Cabinets, autoclave, and drape material are handy for preparing surgical packs and other sterile supplies.
Amy Wolff

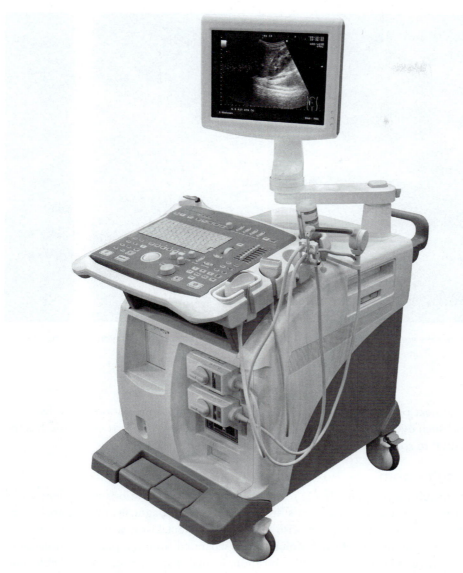

Ultrasound units help provide additional imaging.
DDCoral / Shutterstock

dental radiology equipment, MRI, or CT scan technology may also be among the practice's imaging tools.

Exam Rooms

The exam rooms are where physical exams and client consultations take place. These rooms contain equipment that makes it convenient to conduct a thorough physical exam. This includes a scale, an exam table, an ophthalmoscope/otoscope, thermometers, stethoscopes, and cabinets with supplies. The practice can choose to keep client education materials here as well. Use of the exam room can be for different purposes. One room may be for cats only. Another may be exclusively for sick or injured animals. It is important that exam rooms are clean and odor-free. A staff member should inspect each room after every use to ensure that it is clean

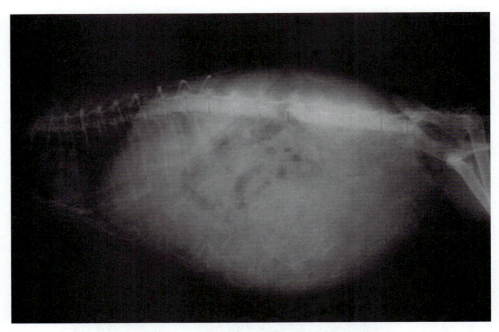

Radiographs are critical to the diagnostic process. How many fetuses can you see in this cat?
VR Photos/ Shutterstock

for the next patient. It is easy to overlook toenail clippings, blood spots, feces, or urine. Such oversight would make a poor impression for your next client and has the potential to spread disease.

Consult Room

The consult room is one that is reserved for private conversations with clients. The team may use it to discuss a pet's medical condition, diagnosis, or prognosis. The doctor or practice manager may wish to discuss a financial issue with a client, and this is best done in an area of the hospital that is separated from the practice's usual busy pace. The consult room is usually a bit more formal than an exam room. Tables and chairs allow a place for the doctor and client to sit down together. The consult room often doubles as a conference room for staff meetings or other business.

Janitorial Room

Every hospital needs a space to store housekeeping items related to building care and maintenance. Brooms, mops, mop buckets, vacuums, and cleaning supplies will need a dedicated space to keep them organized and out of the flow of clinic traffic. It is important to store cleaning supplies in an appropriate manner, to avoid an accidental exposure or toxicity and to distinguish them from preparations intended for patients.

Kennels/Runs

Most veterinary hospitals need to house patients in kennels or runs, depending on their size or medical needs. Kennels are usually arranged in "banks," with a variable number of units, often two or three tiers high. A *run* is a larger containment area, often with a concrete floor and a drain, that allows a larger animal a bit more room and freedom of movement. A large, active dog may require a run for adequate space to accommodate natural postures. A run might also include access to outside areas and may be climate controlled.

Kitchen/Break Room

These areas are a must for a busy staff. Employees need a place to call their own, store and prepare a meal, and relax from their duties. Break rooms may have a table and chairs, microwave, refrigerator, or other small conveniences that make the workplace a bit more like home. There may be lockers or cabinets to store personal items and work-related clothing that is appropriate only for the hospital. These rooms require the cooperation of the entire staff to keep them clean. Resentment builds fast when people do not clean up after themselves. Everyone appreciates a clean place to take a break.

Office

Members of the staff requiring a dedicated work space, computer, library, or other support benefit from an office. The office may be private or shared by several team members. Offices give the hospital team members a space for projects, necessary items, and work-related materials to be collected in one spot. The veterinarian may designate an additional separate space for the doctor's office to work on medical records, return phone calls, or study journal articles and other reference materials.

Bathrooms

Bathrooms (staff and public) must be neat and clean and provide basic necessities. Nobody likes to clean bathrooms, but everyone loves a clean one. It is an expectation of your job to make sure that this area isn't neglected.

Laboratory

The lab will contain necessary equipment to perform "in house" analysis of all kinds of samples. A microscope, blood chemistry analyzers, complete blood count (CBC) machine, centrifuge, and supplies to perform fecal exams, urinalysis, and cytology are all located in the lab.

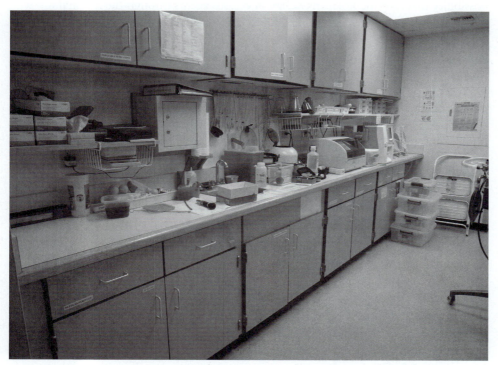

A laboratory area containing blood analyzers, centrifuge, and diagnostic supplies.
Amy Wolff

This hospital keeps a large inventory of pharmacy items.
Amy Wolff

Pharmacy

Drugs and dispensing supplies are kept in the pharmacy area, often closely associated with the laboratory. Some drugs are stocked for "hospital use only," intended only for use on patients while in the practice. Other drugs are for dispensing—medications to be sent home with a client for continued treatment of a pet. The pharmacy may also contain *formularies*. These references list the names, actions, and appropriate doses for prescribed drugs.

Isolation

The isolation area is a special hospital ward where animals with contagious illness are hospitalized or quarantined. This is a preventative measure to keep diseases from spreading throughout the hospital to other patients. Common diseases that might require isolation protocols are canine parvovirus and feline upper respiratory infections. An isolation area requires a strict protocol for entering and exiting the ward, such as donning gowns and gloves and using foot baths or shoe covers. These steps reduce the possibility that staff members may carry infectious agents through the hospital.

Medical Record Storage

Laws require that a practice store all medical records. The length of time varies in accordance with the laws of your state. In a busy practice, the space these records require can become quite large. Medical records not only include the patient's medical folder, but also radiographs or other images that accompany the record. Some practices are now paperless, storing all medical information digitally on hospital computer equipment. This method saves space and is a "greener" alternative to generating a paper record. It also makes record sharing easier, because a practice can rapidly send medical records by e-mail or download them to a CD to make them portable for the client.

Necropsy/Body Storage

Sadly, all hospitals will have to dedicate a space for those patients that die or those they must euthanize. A large freezer to store these patients is a must. Veterinarians must also perform necropsies, examinations of the remains of an animal, to determine the cause of death. Performing a necropsy in a separate area can prevent the spread of disease to patients, and spare the staff and clients exposure to unpleasant odors.

Fenced Yard

Many veterinary hospitals walk patients outside. Some canine patients simply won't relieve themselves if they are confined. A fenced area keeps the patient

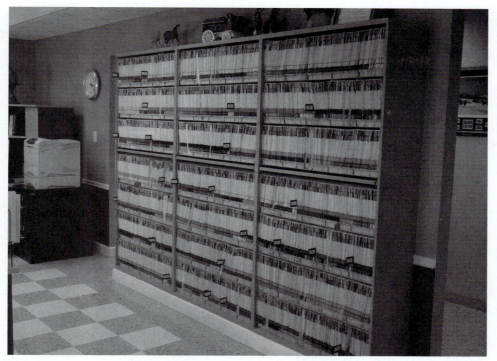

This clinic maintains a well organized system of filing medical records.
Amy Wolff

confined on hospital grounds in the event that they pull away or slip out of a collar and leash. Sanitation is important: you should keep the area clean of debris and feces.

Boarding and Grooming

Boarding provides income for the veterinary hospital. Clients like to board pets with their veterinary facility because they trust the staff to care for their pets. The surroundings and people are familiar to the pets. Some patients with ongoing chronic diseases such as diabetes or congestive heart failure receive better care if boarded in a hospital kennel. The staff can medicate, carefully monitor, and maintain medical records as part of their care while the owner is away.

Grooming is another income source and also promotes good client relations. A pet that is returned to its owner clean and smelling fresh is a sign that the staff cares for the client and the pet's well-being. It conveys the message to the client that you are not just concerned about the pet's medical issues, but also with the animal's overall health.

The task seems daunting! How can Dr. Newton possibly anticipate how to design the hospital work space, equip it, and stock it with the right supplies? This young veterinarian knows how to treat many animal diseases, but no one ever taught him how to design a veterinary hospital. Dr. Newton contacts a veterinary supply company that helps set up new practices. In 24 hours, the company returns a plan to the doctor that lists all his equipment and supply needs, plus provides a diagram of the hospital layout. The company has itemized the costs for Dr. Newton to present to the bank. The company provides this inventory-listing service for free, provided that Dr. Newton buys all his equipment from them. Once his loan is in place, the company will set up a payment plan so that Dr. Newton can afford the new equipment.

With the location selected, the bank loan granted, and the equipment on the way, Dr. Newton now needs a staff. He wants to hire a receptionist, two licensed veterinary technicians, and one veterinary assistant. All these staff members play a vital role in the hospital's success and the patient care.

MEMBERS OF THE VETERINARY TEAM

Veterinarian

A veterinarian is an individual who has graduated from a college of veterinary medicine. The veterinarian can diagnose, treat, prescribe medication, and perform surgery. In order to practice, the veterinarian must be appropriately licensed according to the state's requirements. Veterinarians can practice directly after finishing school. Some elect to go on for further training such as an internship or a residency. Veterinarians can specialize in a particular field and obtain board certification after passing the training requirements and an exam. They can obtain board certification in many disciplines. This indicates that the doctor has reached the highest level of credentialing in that field.

Veterinary Technologist

Veterinary technologists are graduates of a four-year program and earn a Bachelor of Science in Veterinary Technology.

Veterinary Technician

The veterinary technician is a graduate from a two- to three-year program earning an associate degree in Veterinary Technology.

Veterinary Technician Specialist

A veterinary technician specialist has obtained more extensive skills and experience in a certain area. A veterinary technician wishing to specialize in a specific area can do so by contacting the National Association of Veterinary Technicians in America (NAVTA, 2012). Individuals must meet certain requirements within the area of specialization to receive the certification. Dentistry, behavior, anesthesia, equine medicine, emergency, and critical care are areas in which a technician can specialize.

Veterinary Assistant

The veterinary assistant can obtain training through veterinary education programs, but does not receive a degree. He or she receives training at a level that allows him or her to assist the veterinary technician and veterinarian. Some veterinary assistants receive training on the job.

Kennel Attendant

A kennel attendant is responsible for the care and maintenance of the animals that are boarding. The veterinary technicians often supervise kennel attendants and report any changes in health or behavior of any animals left in the hospital's care. Kennel attendants spend a great deal of time and effort maintaining proper sanitation of the building and grounds and practicing good animal husbandry.

Receptionist

It is the receptionist's function to organize the tasks of client service. The receptionist is the "face of the practice" and is the first and last point of contact for the client. The receptionist schedules appointments, answers the phone, maintains medical records, and often takes care of payments and billing.

Practice Manager

A practice manager is a professional office manager specializing in the unique needs of veterinary hospitals. Many practice managers have a bachelor's degree in business management. Veterinarians who wish to be relieved of the many business aspects of running a hospital often hire practice managers to attend to these tasks. Practice managers have the responsibility of hiring and terminations, staff scheduling, training, OSHA compliance, payroll, benefits, marketing, supply ordering, and facilities maintenance.

Groomer

A groomer performs bathing, trimming, and styling of a dog or cat's fur. They restore or maintain a pet's attractive appearance and monitor for general health by careful observation of the skin, teeth, nails, and ears. Clients appreciate picking up a pet after boarding or a hospital stay when the pet looks clean and smells fresh. Many veterinary hospitals won't release a patient to the client without properly grooming it.

Dr. Newton has made a careful study of the components of a well-managed hospital. He has identified all the necessary spaces, as well as the veterinary team members. Because Dr. Newton is opening a new practice without an established clientele, his needs at this time are small. He is planning on hiring one full-time receptionist, one full-time veterinary technician, and two part-time veterinary assistants.

PRACTICE DOCUMENTS PROVIDE A SOUND FOUNDATION

Before the hiring process begins, Dr. Newton must have some important documents in place. They will serve as the foundation for appropriate business practices and keep him in compliance with federal and state laws as an employer and a practice owner. He will have to spend some time organizing these documents and may even consult a veterinary practice manager for an action plan. Taking all these steps before employees are hired makes sense. To hire staff without them would be like holding auditions for a play and not being able to tell people what roles you are casting.

Below is a list of documents that help keep clinics organized and on track.

Job Descriptions

A job description provides detailed information about the responsibilities of each position in the hospital. It will describe the various tasks of the job, the necessary skill sets, and the educational requirements for accomplishing these functions. The job description may also set performance goals to ensure that employees meet these expectations.

Staff Training Manual

Writing a staff training manual is a big job. The manual will contain descriptions of hospital policies and procedures covering all aspects of employment. The manual provides clarifications and expectations of employee performance so that the employer can measure it. Every hospital will have a training manual that addresses its specific needs, but some of the more common policies include scheduling, overtime

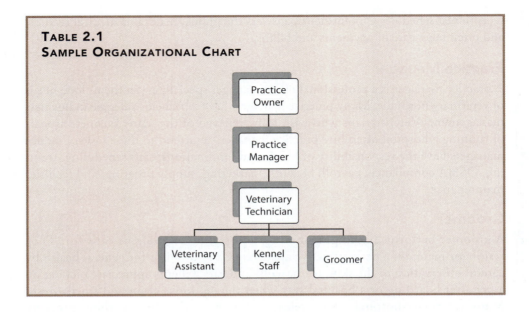

TABLE 2.1
SAMPLE ORGANIZATIONAL CHART

pay, benefits, employee discounts, sick days and medical leave policy, continuing education, dress code, smoking policy, workplace safety, housekeeping, and professional conduct.

Training Manual Topics

Organizational Chart   The staff training manual will include information on the organization of the practice—sometimes referred to as the "org chart"—so that employees know whom to report to as their direct supervisor. This helps establish a chain of command and decreases the chances of multiple people trying to solve the same problem. See Table 2.1 for a sample organizational chart.

OSHA Protocols   The Occupational Safety and Health Administration is a government agency that regulates workplace health and safety. The Occupational Safety and Health Act of 1970 requires that each employer meet the safety and health standards designed to provide a safe workplace. OSHA training allows employees to become familiar with these safe workplace habits. Training usually takes place once a year, and your employer will document this training in your personnel file. OSHA training covers topics that you need to be familiar with, regarding safety in the workplace, how to reduce or eliminate the risks of exposure to hazardous chemicals, and even how to lift properly to avoid back injuries. OSHA will require appropriate signage in the hospital to identify exits, floor plans, and evacuation routes in case of an emergency. It is important that a hospital is OSHA compliant, and usually the practice designates one staff member to make sure that the entire staff participates in employee training and that all employees follow other OSHA protocols (OSHA, 2012).

Material Safety Data Sheets (MSDS)   A Material Safety Data Sheet is a document that provides important information about the composition and the physical properties of a chemical product, as well as information regarding toxic exposures, storage, disposal, spill handling procedures, and any required protective equipment. The manufacturers and distributors of these products are responsible for supplying this information. They may enclose an informational sheet with the product when it is shipped or offer a website where you can access and print sheets as needed. For example, in preparing a solution of disinfectant for mopping the isolation area, you accidentally spill some of the concentrated stock solution on the floor. The MSDS will provide the information necessary for you to handle the spill properly, without exposure to a harmful chemical. MSDS sheets are usually kept in a binder, and every employee must know its location so that he or she can readily access it in case of emergency.

Every staff member must know how to appropriately handle biohazard material.
Travis Klein / Shutterstock

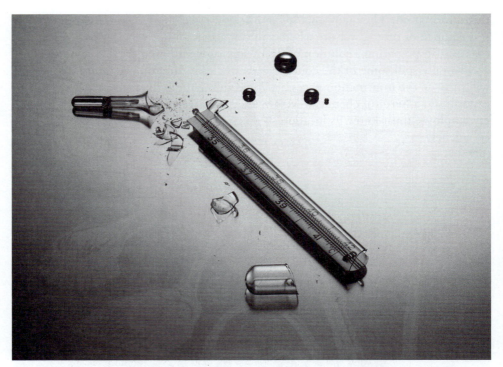

Mercury from a broken thermometer is a hazardous spill that must be cleaned up properly.
AntonioFoto / Shutterstock

Safety Protocols/Manuals   What would you do if a bottle of gas anesthetic fell and broke in the treatment area? How would you handle an aggressive dog that is lunging at the door to the kennel when you approach? A hospital safety protocol will outline the steps you must take for anticipated situations in the veterinary hospital. You cannot foresee every problem that may arise, but you can anticipate many of the common safety issues. A *safety manual* consists of all the various protocols. It often has an index, so if you do drop the bottle of anesthetic, you could look up "hazardous spills" and find instructions on how to appropriately take care of the problem.

Biohazard Protocols   A biohazard is an infectious agent or material that presents a potential risk to the health of humans, animals, or the environment. You must be careful to dispose of biohazards in an acceptable manner, to avoid spreading an illness or to prevent an injury. Examples of biohazards include used needles and syringes, bacterial and fungal cultures, and also the fecal waste from patients with contagious diseases such as canine Parvovirus.

Hospital Policies and Procedures   Every hospital will establish policies and procedures for both the business and medical side of the practice. For example, if you have a client who is unable to afford treatment, will your hospital extend credit, or are all fees due at the time that you discharge the patient? Are your clients allowed into the treatment areas? Is a client permitted to restrain his or her pet during its physical exam? All veterinary hospitals must make decisions on questions such as these so that resulting established policies can reduce the liability and risk of caring for patients. For example, did you know that if a client is bitten by his or her own pet while on the premises of the veterinary hospital, the client can sue the hospital? To avoid this potential situation, the veterinarian may insist that only team members restrain all animals. The practice develops policies and procedures to avoid situations that are detrimental to the practice.

Chris Pole / Shutterstock

Staff Performance Evaluations   Dr. Newton wants to make sure that he can monitor his staff's performance by conducting evaluations. He can do this once or twice a year or after a probationary period in the case of a newly hired team member. Performance evaluations will be most effective if Dr. Newton gives each staff member defined and measurable goals. These can include accomplishments that enhance patient care or increase the income of the practice. For example, when you start your new job, one of your goals might be to master and demonstrate proficiency using the blood chemistry analyzer. When Dr. Newton conducts your performance review, you'll be able to show him that you can conduct this task without hesitation or supervision. You have achieved your goal (and maybe a raise!). On the other hand, if you have not achieved the expectation by the time you have your evaluation, it may signal to Dr. Newton that you have not been an effective employee. He may ask you to attend special training or to spend more time practicing a skill, or he may even relieve you of duty. Performance evaluations help define clear goals and a time frame within which the employer expects you to meet them. Job descriptions should contain measurable

alexskopje / Shutterstock

goals so you can see if you are meeting your job expectations. For example, one of the expectations of your job might be to increase the number of heartworm exams by 20 percent in the next six months.

Hiring and Firing Procedures   When you land your first job, when can you expect a raise? Do you receive health benefits? What types of situations can result in your immediate dismissal from your job? Hiring and firing policies define what to anticipate when starting your new position and what actions can result in the loss of your job. Most hospitals will expect you to complete some employee training or have a probationary period before you are awarded full staff benefits. They might ask you to sign a confidentiality agreement, a noncompete contract, or a statement indicating that you have received employee manuals or policies and procedures handbooks. Hospitals will set their own policies about what actions will result in immediate dismissal, but theft of hospital property, coming to work altered from drugs or alcohol, insubordination, failing to come to work without notification, or patient/client or coworker abuse are almost always included.

Salaries/Benefits/Bonuses/Profit Sharing   Dr. Newton must decide on a pay scale for his employees. Because the doctor understands the value of having trained veterinary personnel, he has decided to offer a higher salary to his licensed or credentialed professional staff, with raises for every benchmark documented on their performance reviews. Right now, because the hospital hasn't produced any income, he will only be able to offer vacation and sick leave benefits. He hopes to add medical insurance, 401(k), and continuing education expenses at a later date. A bonus is a good way to reward staff performance without having to give a raise. Bonuses might be money, events, or gifts. Profit sharing means that the staff has a vested interest in the practice. The more money the hospital makes, the bigger the percentage that staff gets. This serves as an incentive to ensure that the hospital charges appropriately for services. It is easy in the course of the day to give away small services that amount to revenue. Just think: if you didn't charge for one $10 nail trim every week, your practice would lose $520 this year, which just might be your performance bonus.

rasskazov / Shutterstock

It is standard practice to have an employee sign an acknowledgement form indicating that he or she has received a copy of the training manual. This protects your employer. It prevents an employee from claiming that he or she was not properly informed of a hospital policy if a violation occurs.

Things at All Pets Animal Hospital are beginning to look up. However, there are still some important tasks left to complete. Before All Pets can open, Dr. Newton needs to make sure he has obtained the following:

Malpractice Insurance

Malpractice insurance provides financial coverage in the event that somebody sues Dr. Newton for malpractice. Malpractice is the act of causing damage or injury as a result of negligence. In our society, people will often bring suit if they feel a wrong treatment, diagnosis, or outcome has occurred. Dr. Newton doesn't *think* he needs this type of insurance coverage, but he's been getting magazines and publications with case reports of clients who have sued their veterinarian for all kinds of mishaps and misunderstandings. Our doctor can purchase malpractice insurance from a private carrier or from the American Veterinary Medical Association (AVMA PLIT, 2013).

Workers' Compensation Insurance

Workers' compensation insurance provides medical care and wages for employees who are injured while performing duties at work. In exchange for these benefits, the employee forfeits the right to sue the employer for the work injury. This type of insurance is required by law and controlled by the state government (U.S. DOL, 2012).

Accounting and Tax Advice

Dr. Newton took nothing but science classes in school. He can barely balance his checkbook. He will need an accountant to manage his payroll, rent, bills, and other

Dusit / Shutterstock

expenses and to make sure he isn't overspending. He will need help filing taxes as well, because many of the expenses he will incur to start this practice may be tax deductible. He wants to make sure that he is managing his money correctly and knows that an accountant will accurately calculate his taxes and see that the taxes are paid on time.

Inventory Control System

When All Pets Animal Hospital officially opens, the hospital will start out with a limited amount of drugs and supplies. This prevents Dr. Newton from tying up too much money in products that sit on the shelf. Until his hospital has a dependable client base, he risks stocking items that might expire before he can use them. He needs a cash flow right now to meet the hospital's expenses. As the practice grows, the practice manager will begin to formulate an inventory list that includes every item in the hospital and will program it into the computer. Dr. Newton is purchasing hospital management software that has a tracking system to keep a record of all sales from drugs and other inventory. This way, he can tell what items he uses and sells most frequently, how many to have on hand, and when it is time to reorder.

Forms, Forms, and More Forms

In order to have the proper documentation for a patient's medical record, our hospital will have to design many forms. There will be consent forms for procedures, treatment forms, physical exam forms, discharge instruction forms, and lab forms, just to name a few. Standard word processing software can generate most of the forms that a hospital requires. The hospital can customize forms with the hospital logo, address, phone numbers, and staff names and even personalize them for the patient with pictures and other graphics.

Marketing and Advertising

Last, but not least, Dr. Newton wants to get the practice off to a good start by making sure that he *advertises*. He wants the public to know where his practice is located, that his primary focus is small animal medicine, and that he is open for business. Placing ads in local papers and the phone directory, and even doing a radio announcement, are examples of advertising. Advertising is a part of *marketing*. Marketing a practice refers to all of the steps Dr. Newton will take to bring people to All Pets Animal Hospital and turn them into clients. The doctor knows that word of mouth is great advertising, and he plans on delivering high-quality medicine and service. However, there are other ways to advertise the practice that Dr. Newton plans to use. He will hire a website designer so that his clients can keep track of hospital news online, request medication refills, and even book appointments. Social networking is vital, and he is planning on using many of the available social media to talk about his practice and the hospital services.

Marketing is crucial because it helps to inform the pet-owning public about animal health care needs and Dr. Newton's desire to meet those needs. Not every owner knows the benefits of dental care, annual vaccinations, and nutrition. Annual events such as Dental Health Month, holding puppy socialization classes, and speaking to groups in the community about pet health care topics are ways to gain exposure for the clinic and promote client visits. Having the logo or name of the hospital on leashes, calendars, pet bandanas, and other small thank-you tokens advertises and markets the hospital. There are many ways to market All Pets Animal Hospital. Below are some common marketing strategies:

1. Hold seminars about pertinent health topics at your clinic, and invite the public to attend.
2. Reward existing clients for referring new clients to the practice with a small gift or discount on pet services. Increase the value of the thank-you for each successive client referral.
3. Participate in community events with a booth and a small packet of information, along with a pet treat. Have separate bags for dogs and cats.
4. Dress the employees as a team, with the hospital name and logo. Lightweight jackets, T-shirts, and sweatshirts get the name of the hospital out in public.
5. Offer special services such as house calls, in-home euthanasia services, or hospice care.
6. Volunteer at local shelters and rescues. Demonstrate how the caring staff of All Pets Animal Hospital takes the time to help these organizations.
7. Talk to school-age children about pet care. Make it age appropriate. Bring a pet that is outgoing and friendly as a goodwill ambassador.
8. Send expressions of sympathy to clients over the loss of a pet. A card or small floral arrangement sent on behalf of the entire staff tells the client that the staff feels the loss of the pet too.
9. Produce a hospital newsletter or a blog that spotlights a staff member or an interesting case of the month (with permission from the owner!). Discuss a common medical condition or an emerging disease.
10. Use the staff wisely and effectively to provide customer service and patient care that keeps the client coming back.

Wearing the same T-shirts identifies team members.
Amy Wolff

CLIENT CARE

Communication and Customer Service

Dr. Newton has really done his homework. He is building a strong foundation and a sound business plan for All Pets Animal Hospital. He plans on hiring a professional staff with a strong work ethic who will be accountable for providing a safe work environment, implementing clearly stated policies and procedures, and providing excellent patient medical care. At the same time, Dr. Newton must reinforce to the staff the importance of caring for the *client*. The client is the person holding the leash, the cat carrier, or the halter, or standing at the end of the chute. If the client is dissatisfied with the service at All Pets Animal Hospital, it won't matter how smart Dr. Newton and the staff are. The clients will seek service at another hospital. How will the employees at All Pets Animal Hospital treat their clients? The cornerstone is *communication.* More problems and conflicts arise from poor communication than any other reason. You must greet every client in a warm and caring manner. Treat every client with dignity and respect. Promptly inform him or her of any issue that arises during the pet's visit or care. Inform clients if the doctor is running behind. Provide a clear explanation and estimate of the cost of care. The staff should be well groomed and look professional, to indicate to the client that they maintain a high standard of performance. The hospital should always be clean and free of odors. The outside of the building should reflect these professional values as well. If you care for your clients, they will trust Dr. Newton and his staff to care for their pet . . . and they will tell their friends.

Flawless customer service should always be your goal.
iQoncept / Shutterstock

SUMMARY

The veterinary team comprises educated, skilled professionals who work as a unit to provide health care for animals.

Establishing a veterinary hospital is a complex business. From start to finish, there are many details to consider in the design, location, staffing, and equipping of a veterinary clinic. A good foundation starts with a mission statement, so that the veterinary team understands the purpose behind the practice and can assess whether or not they are meeting their goals. A solid business plan and clear hospital policies and procedures guide the team in complying with local, state, and federal laws. The hospital needs a sound management foundation in order to succeed.

There are many types of veterinary facilities serving a wide variety of patient needs. A veterinary hospital may offer service to only one or all species of animal. The practice might be limited to a specific type of service, such as emergency cases or surgical referrals. Often, the veterinarian's training and preferences determine the type of practice. Appropriate marketing and advertising help the practice remain an integral part of the community.

TEST YOUR KNOWLEDGE

1. Explain the purpose of a mission statement as it relates to the practice of veterinary medicine.
2. Discuss the differences in building design and location between an equine practice and a small animal hospital.
3. List accessories or "creature comforts" that you could make available in your reception/waiting area to make the clients and the patients more comfortable.

4. Explain how a referral center differs from a veterinary hospital. Identify the type of patients that you might send to a referral hospital.

5. Compare and contrast the difference in educational requirements between a veterinarian, a veterinary technologist, a veterinary technician, and a veterinary technician specialist.

6. Summarize the process of planning a new veterinary practice, starting with the business plan and ending with the opening of the hospital.

7. Which of the following tasks would fall under the Occupational and Safety Health Act of 1970?
   a. The ordering and storage of controlled drugs
   b. The use of eye protection when working with hazardous chemicals
   c. The type of flooring that is used in the hospital ward

   d. How often an employee is allowed to take a work break

8. Mr. Richards has just completed his visit with his three dogs for their annual exams and vaccinations. At checkout, he asks if he can have a multipet discount. What resource in the hospital will have the information to answer Mr. Richards's request for a discount?

9. Compare and contrast *marketing* and *advertising* in terms of promoting a veterinary practice. What marketing strategies do you think would help bring clients into a veterinary practice?

10. Mrs. Schmidt is very upset about her bill. She was not planning on spending anywhere near the invoiced amount for her dog Summit's neuter surgery. What steps could she have taken to avoid this situation?

BIBLIOGRAPHY

AVMAPLIT. 2013. "Home Page." Accessed May 1, 2012 from the American Veterinary Medical Association Professional Liability Insurance Trust website.

DEA. 2012. "Registration Home Page." Accessed June 30, 2012 from the United States Drug Enforcement Administration website.

NAVTA. 2012. "Committee on Veterinary Technician Specialties." Accessed June 21, 2012 from the National Association of Veterinary Technicians in America website.

NBVME. 2012. "Home Page." Accessed July 17, 2012 from the National Board of Veterinary Medical Examiners' website.

OSHA. 2012. "Home Page." Accessed July 18, 2012 from the Occupational Health and Safety Administration website.

U.S. DOL. 2012. "Workman's Compensation." Accessed July 18, 2012 from the United States Department of Labor website.

SBA. "Home Page." Accessed July 18, 2012 from the U.S. Small Business Administration website.

3

Legal and Ethical Issues in Veterinary Medicine

Learning Objectives

At the end of this chapter, you should be able to:

- Explain the difference between legal and ethical issues.
- Identify the government agencies and national associations that make and enforce laws for the practice of veterinary medicine.
- Compare and contrast the difference between legislative law and common law and give an example of both.
- Identify and discuss current ethical issues in veterinary medicine.

INTRODUCTION

As a veterinary professional, you begin your career with the expectation of caring for the health of animals. You are now actively engaged in learning the *science* of medicine and the *principles* of health, husbandry, and disease. You may not have realized that an important aspect of your job will be to follow the laws, ethical principles, and professional expectations that are the foundation for veterinary practice. What are these laws and ethics? How do you tell the difference between a legal and an ethical issue? What impact do laws and ethics have on the practice of veterinary medicine? As a veterinary technician, how do these laws affect your ability to provide care for animals? You will identify the differences between legal and ethical matters as we explore examples and scenarios of these issues.

WHAT IS A LAW?

Legislative Law

Simply stated, laws are sets of rules that regulate our behavior or actions. They provide a framework for a stable society, so that each member has an expectation of what is legal and illegal, and what punishment we may expect if we break the law. We refer to laws that the government has written, approved, and enforces as *legislative laws.*

Laws make a community safer and cleaner and serve as a platform for justice. Authorities whom we elect or appoint enforce laws. Counties, cities, states, or the federal government can set these laws. If you break a law, you must answer for your behavior and face the consequences.

You are familiar with all kinds of legislative law. Speed limits determine how fast you can drive. You cannot take merchandise from a store without payment. You cannot buy alcohol or tobacco if you are underage. These are examples of legislative laws that affect all citizens.

Common Law

In contrast, common law refers to decisions judges make through court cases. These cases come to trial when the respective parties cannot resolve a disagreement. There may be harm caused to a person or property, or someone may have represented him or herself dishonestly. The judge must decide which party wins the dispute. If two neighbors are arguing over responsibility for the care of a tree that straddles their property lines, a judge would make a decision in a civil court based on *precedence.* Precedence means that a judge bases the decision on prior, similar resolved court cases.

LEGISLATIVE LAW PERTAINING TO VETERINARY PRACTICE

Laws that pertain to the practice of veterinary medicine are both legislative (written down) and common (civil matters decided by a judge). You should understand these laws, as they govern the way that we practice medicine, care for our patients, and interact with our clients. They ensure that we provide the best quality of care.

Veterinary Medical Practice Act

Each state sets the laws and regulations that govern the practice of veterinary medicine. In most but not all states, we refer to those laws as a Veterinary Medical Practice Act. A Veterinary Medical Practice Act (VMPA) is a set of rules, regulations,

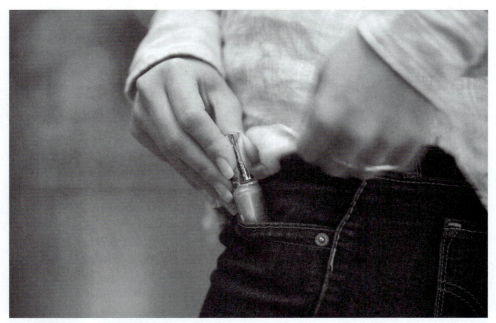

Lisa S. / Shutterstock

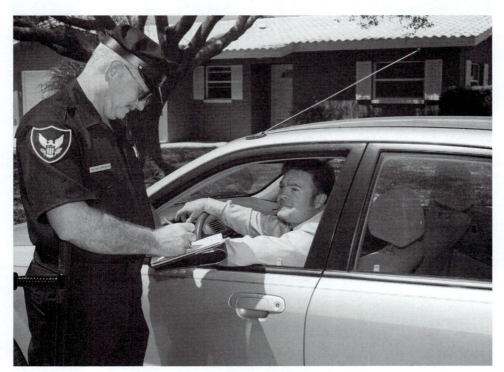

Shoplifting and speeding are prohibited by legislative laws.
Lisa F. Young / Fotolia LLC

and professional principles that address the relationship and obligations between veterinary staff and the client. Remember, veterinary health professionals are responsible for public health, safety, and welfare, as well as the care of our animal patients. There must be rules in place to ensure that we conscientiously follow these principles. Each state has the freedom to draft its own individual set of rules, but three of the profession's leading organizations have written model practice acts that

© Junial Enterprises/Fotolia

serve as a template. These model templates serve to make the practice of veterinary medicine as consistent as possible from state to state. (See the model practice acts at AVMA, NAVTA, and AAVSB websites for more information). The NAVTA model practice act is structured specifically to address the role of the veterinary technician (NAVTA, 2009).

To enforce the rules and regulations, each state will have a regulatory body, which is frequently a veterinary medical board. Board members consist primarily of veterinarians, but may also have a veterinary technician or representative from the general public. The board members have the responsibility of running the affairs of the practice of veterinary medicine at the state level. This includes upholding the Practice Act, hearing cases of medical malpractice and negligence, issuing facilities permits and disciplining, and suspending or revoking licenses of veterinarians and veterinary technicians who are in violation of state law.

A Practice Act addresses the rules and regulations regarding the practice of veterinary medicine. Among them are the laws governing veterinary technicians. Let's review a few parts of the model practice act from the AVMA that pertain to your job as a veterinary technician.

1. A Practice Act regulates the licensing and credentialing of veterinary technicians. It states the qualifications a veterinary technician must have in order to be licensed. There is considerable variation across the country pertaining to the licensing of veterinary technicians. Most states require the veterinary technician to be a graduate of an AVMA-accredited school and successfully pass both the Veterinary Technician National Exam (VTNE) and the state exam before he or she can obtain a license. Currently, only Alaska, Delaware, Wisconsin, and Washington recognize on-the-job training and alternative pathways. All other states require veterinary technician applicants to graduate from an AVMA-accredited program.

2. A Practice Act may state the requirements for continuing education, if needed, in order for you to keep your license active. You may be asked to retain proof of your attendance or other verification to send in with your annual license renewal.

3. A Practice Act may define the tasks a veterinary technician can legally perform and how much supervision is required while performing those duties. We refer to these as *levels of supervision*, more fully discussed in Chapter 1.

4. A Practice Act describes and defines the Veterinary-Client-Patient Relationship (VCPR). The VCPR infers that the veterinarian has taken responsibility for making decisions about the patient's health and treatment and that the client has agreed to follow the recommendations. It also states that the veterinarian must have examined the patient in person and has sufficient knowledge about the patient's condition to make an initial diagnosis. For example, if a new client sends a video clip of his dog coughing through clinic e-mail, this would not qualify as a valid VCPR. A veterinarian cannot prescribe drugs for a patient that he or she has never examined. The VCPR also requires that the veterinarian inform the client how his or her pet should receive follow-up care or emergency services. It is not a requirement for the veterinary hospital to provide emergency services, only to inform the client where he or she may obtain help should the need arise.

Although the VCPR primarily affects the veterinarian, the entire veterinary staff is responsible for upholding the conditions of this relationship by maintaining detailed medical records including test results, laboratory information, prescriptions, and phone conversations with the client. This information is a legal document, and it is the responsibility of the entire staff to make sure it is complete, accurate, and up to date. We will discuss the medical record in more detail in Chapter 6.

Laws Regarding Safety in the Workplace

You are now familiar with laws that regulate the qualifications of veterinary professionals and the standard of care that they provide for patients. Do you know that there are laws that protect *your* safety in the workplace? Caring for animals will expose you to numerous workplace hazards that can endanger your health and safety.

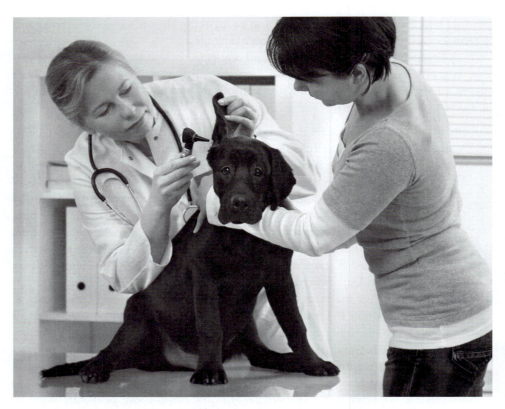

A valid Veterinary-Client-Patient Relationship must exist for a veterinarian to make any decisions about a pet's health care.
Alexander Raths / Shutterstock

By federal law, all employees are entitled to a workplace that uses approved measures to eliminate or reduce your risk of injury at work.

OSHA and NIOSH   In 1970, The United States Department of Labor drafted the Occupational Safety and Health Act. The primary objective of this legislative law was to set standards that employers must meet to reduce your exposure to workplace hazards by enforcing safety requirements and decreasing workplace hazards. Two agencies were created to accomplish these goals, The Occupational Safety and Health Administration (OSHA) and The National Institute for Occupational Safety and Health (NIOSH). OSHA developed and enforces the workplace laws. NIOSH provides research, education, and training in the field of workplace safety. Most veterinary technology students don't give much thought to the number of potentially hazardous situations that are commonplace in veterinary medicine. Below is a list of health risks that you may encounter working in a veterinary practice.

1. **Exposure to hazardous materials**
   a. **Anesthetic waste gas.** There is waste gas produced when animals are anesthetized with inhalant anesthetics. There must be measures in place to reduce or eliminate your chances of inhaling the fumes that do not go directly into the patient's respiratory system. The use of NIOSH masks, activated charcoal canisters, and hospital scavenging systems reduce these exposures.
   b. **Compressed gasses.** Medical cylinders that contain oxygen (O_2) or nitrous oxide (NO^2) require safe handling and storage procedures to minimize the risk of explosion and combustion. The use of eye protection, rubber mallets, and properly stored tanks chained to the wall minimizes risks.

A NIOSH mask protects you from inhaling waste anesthetic gases.
Adem Demir / Shutterstock

The F-Air canister utilizes charcoal to absorb anesthetic waste gases.
Amy Wolff

Stored oxygen cylinders chained to the wall.
Amy Wolff

c. **Chemotherapeutic agents.** Drugs used to treat cancer in animals require personal protective equipment and safe disposal methods to ensure that veterinary personnel do not come into contact with these agents or discharge them into the environment. Personnel delivering these drugs should use gloves, gowns, and masks. An approved spill kit should be on the premises in case of an accident.

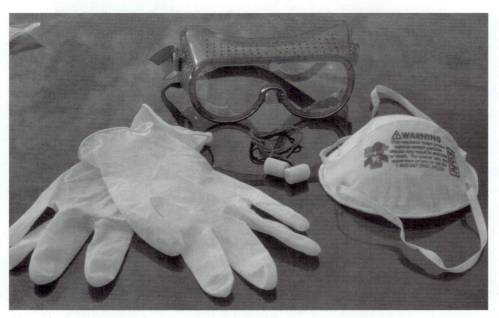

Personal protective equipment is essential when working with biohazardous material.
Rob Byron / Shutterstock

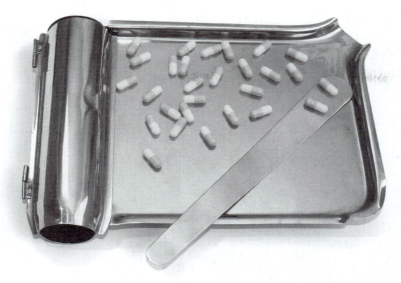

Praisaeng / Shutterstock

d. **Pharmaceuticals.** You will handle a variety of different drugs on a daily basis. You must take caution to avoid exposure. Use tools such as counting trays and spatulas in place of direct handling to ensure that you do not come into contact with prescription drugs. This avoids the potential of absorbing drugs through the skin or provoking allergic reactions. There are some pharmaceuticals that you should only apply when wearing gloves.

e. **Tissue fixatives.** Veterinarians and veterinary technicians often use formalin as a tissue preservative for specimens that a pathologist will then examine. You must take care to avoid skin contact and breathing the fumes. Use gloves and a NIOSH mask, and make sure ventilation is adequate.

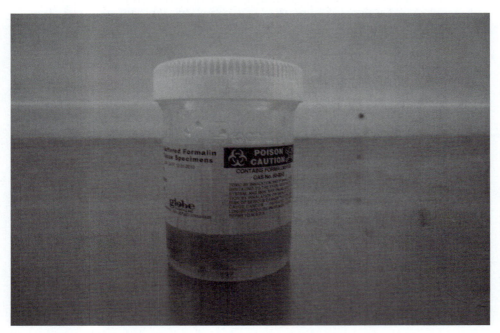

Notice that the label indicates that tissue fixative (formalin) is a biohazard.
Amy Wolff

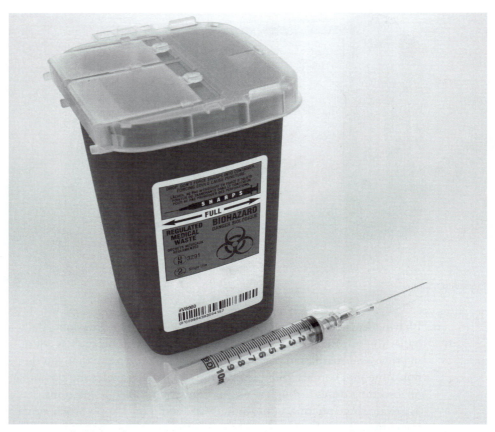

Dispose of sharps and other medical waste in approved containers only.
magicoven / Shutterstock

    f. Pesticides. Both large and small animal practices routinely use pesticides to control insects and other ectoparasites. These are used in the building and on the grounds, as well as on livestock and companion animals. Use gloves, NIOSH mask, aprons, boots, and safety glasses.

    g. Medical waste. Blood, tissue, and body fluids can potentially carry disease organisms. You must handle and dispose of them properly to avoid personal contamination and the introduction of disease into the community. It might surprise you to know that exposure to human blood and body fluids sometimes occurs. Injuries can arise when clients try to transport a pet to the clinic or restrain their own pet. Universal precautions include exam gloves, safety goggles, face masks, and disposable gowns. Use proper medical waste disposal containers and biohazard bags, and locate them conveniently throughout the practice.

    h. Chemicals/miscellaneous. These include disinfectants, cleaning agents, latex, and surgical antiseptic scrubs. Use gloves, eye protection, boots, and NIOSH masks to avoid exposure.

  2. Exposure to diagnostic radiation

    a. X-Rays. X-rays are electromagnetic energy that can cause damage to living cells. You are *required* to use personal protective equipment (PPE) to protect you from excessive exposure to hazardous radiation. Lead lined gowns and gloves, a thyroid protector, and eye shields are designed to minimize X-ray exposure to sensitive organ systems. You will also need to wear a

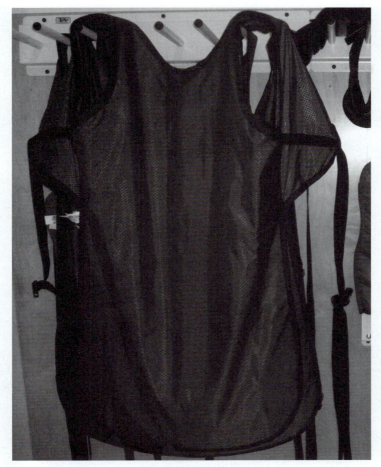

Amy Wolff

dosimetry badge that monitors and measures the amount of exposure that you have had to diagnostic radiation within a pre-determined time period.

b. X-ray fixative/development solutions. These can irritate the respiratory tract and cause skin reactions in sensitive individuals. Use gloves, aprons, and eye protection when working with these chemicals.

3. Exposure to damaging light

Lasers. The use of the laser has found many applications in the veterinary hospital. Practices use carbon dioxide (CO_2) lasers for surgical procedures; these replace the need for the scalpel by creating a precisely placed incision that is free from hemorrhage. Surgical lasers produce concentrated light that requires you to use appropriate eye protection while the laser is in use. The laser safety glasses contain a filter to protect your eyes from damaging wavelengths. The lasers also produce a mild smoke from the tissue as it cuts. A vacuum hose diverts this smoke from the surgical field. Therapy lasers work to diminish pain and inflammation in damaged tissue, improving return to function after surgery or injury. You are required to use eye protection while using this laser on your patients.

4. Exposure to physical injury

a. Lifting. Lifting patients, cartons of supplies, bags of food, and other items is a way of life in veterinary practice. All veterinary personnel must learn the techniques of proper lifting in order to avoid back

Specialized eye wear will provide protection from ionizing radiation and laser light.
lightpoet / Shutterstock

injury, strains, and sprains. (This includes asking for help.) Devices and supports to protect knees and the lower back can help you avoid injury. Please don't minimize the amount of pain and disability that can result from poorly executed lifting or overestimating the amount of weight that you can manage by yourself. Physical injury can accumulate over the life of your career, and you can find yourself managing chronic pain.

b. Restraint. Proper restraint of veterinary patients often requires you to assume all manner of body positions, both on the floor and on elevated surfaces. Large animal technicians will deal with patients many times their size. Attempted restraint of uncooperative or fractious patients five pounds to five hundred pounds can result in injury.

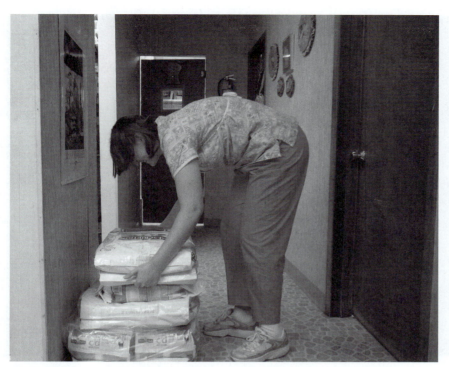

Bent at the waist, this type of lifting will strain or injure your back.
Amy Wolff

    c. Animal bites, scratches, and kicks. Almost all veterinary personnel are affected by this aspect of practice life. You will employ equipment and effective restraint techniques, including chemical sedation, to reduce the incidence of injury. It is important to master the proper use of restraint

Lifting with the knees saves your back from strain and injury.
Amy Wolff

Notice how this technician has control of the patient with the use of a muzzle and proper restraint. She is using her elbow to control the dog's hindquarters, and the dog's head is in the crook of her arm. Her face is a safe distance away from the animal's mouth.
Amy Wolff

poles, gloves, muzzles, cat grabbers, humane traps, leashes, and blankets. You will also need to learn minimal restraint techniques that decrease patient anxiety and stress.

d. Burns: autoclave/electrocautery. There are hot surfaces in the hospital. Autoclave temperatures reach close to 300°F and vent steam. Some

Use of a soft muzzle is often necessary for patients that are at risk to bite.
Gina Callaway / Shutterstock

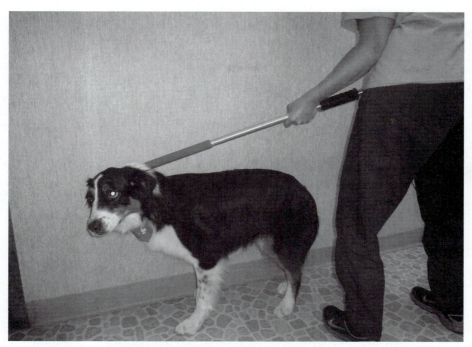

A restraint pole is used to ensure that a safe distance is maintained between you and an aggressive patient.
Amy Wolff

autoclaves are designed to use ethylene oxide gas to sterilize items that are unable to tolerate high temperatures. The gas is toxic and requires proper ventilation. Electrocautery supplies that the veterinarian uses in surgical procedures produce radio waves to create heat that coagulates blood vessels and tissue. Use oven mitts for the autoclaved supplies, and ventilation and a respirator for ethylene oxide gas.

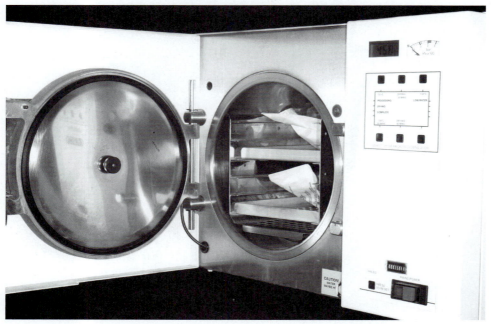

An autoclave is used to sterilize instruments and other supplies. Care must be taken to avoid exposure to steam and hot surfaces.
Chris Pole/ Shutterstock

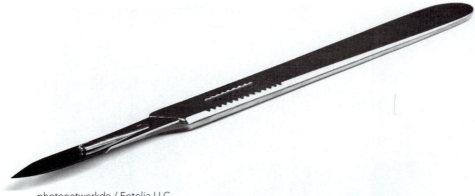

photonetworkde / Fotolia LLC

e. **Sharps: needles and scalpel blades.** Vaccinations, injectable medications, catheter procedures, and surgery all utilize needles and scalpel blades that can cause punctures and cuts if you do not handle and dispose of them properly. Isolate all sharps from a surgical pack; use scalpel blade removers and proper medical waste containers.

5. **Exposure to zoonotic diseases and parasites.** All veterinary personnel must be vigilant about protecting themselves against diseases that spread from animals to humans. Many of these diseases may be only a nuisance; others are life threatening. You must be familiar with the risks that you encounter from your animal patients, as they can impact your own health. Following are a few examples of diseases that you can catch from veterinary patients.

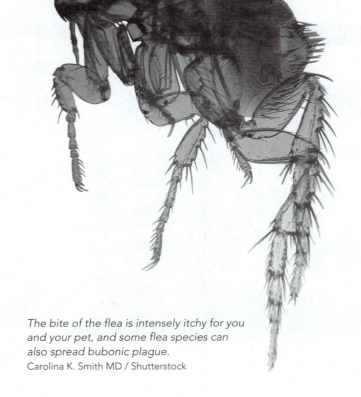

The bite of the flea is intensely itchy for you and your pet, and some flea species can also spread bubonic plague.
Carolina K. Smith MD / Shutterstock

verdateo / Fotolia

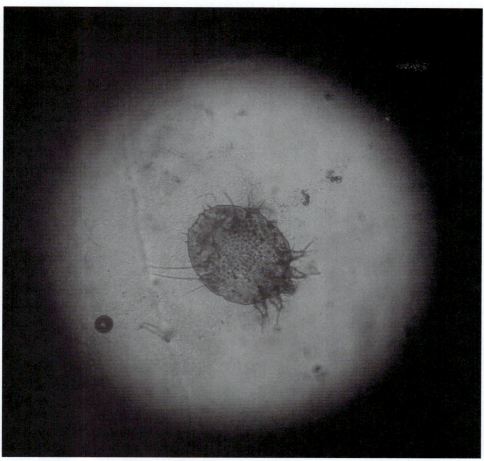

Sarcoptic mites can be transmitted to people.
molekuul.be / Shutterstock

a. Viral: Rabies

b. Bacterial: *E. coli* spp., *Salmonella* spp., *Clostridia* spp., *Bartonella* (cat scratch disease), *Yersinia* (plague), *Psitticosis* (parrot fever), *Brucella abortus* (brucellosis)

c. Fungal: Dermatophytes (ringworm)

D. Kucharski K. Kucharska / Shutterstock

    d. Parasitic: Hookworms, roundworms, fleas, mites

    e. Tick-borne pathogens:* Rocky Mountain spotted fever, Lyme disease, ehrlichiosis

Follow these simple measures to decrease your chances of contracting or spreading a zoonotic disease.

Eat/drink only in designated areas, usually the employee break room. Don't leave your lunch in the lab. It is not smart to place your food and drink items in the same place where feces and urine specimens are processed.

Wash your hands frequently. If you can't remember the last time you washed your hands, it's time to wash them. It may sound excessive, but try to wash your hands before and after every patient, before and after you use the restroom, and before you eat.

Use scrubs/shoes for the workplace; change before you go home. You don't want to be carrying home on your clothes an infectious disease to your own pet. Leave your work clothes at work, or change clothes and shoes before you leave the workplace.

Use PPE (personal protective equipment). Don't get lazy about using the personal protective equipment required to keep you safe. Gloves, goggles, and masks can decrease your exposure to pathogens.

Employers have additional responsibilities for maintaining workplace safety: OSHA standards require your employer to comply with many regulations. Proper safety equipment must be available. Your employer must keep records of workplace injuries and illnesses, and provide educational materials that teach you about safety in the workplace and how to appropriately respond to accidents, injuries, and emergencies. Employers are also responsible for finding and correcting safety hazards by changing the work environment and minimizing the risk.

1. **All clinics must have an official OSHA poster displayed, informing employees of their workplace rights.** This poster provides detailed information concerning your right to a safe working environment and numbers to call if you wish to report a violation that your employer has not addressed.

2. **Maintain information about every drug and chemical that the clinic uses.** The Material Safety Data Sheets (MSDS) that accompany drugs and chemicals provide detailed information about the chemical structure, possible hazards, and proper procedures for handling spills and contamination. All businesses, including veterinary clinics whose employees work with these hazards, must have an MSDS database.

3. **Provide yearly training regarding workplace hazards, personal protective equipment, and emergency protocols.** All veterinary personnel must go through yearly training in workplace safety, including a review of the emergency evacuation plan, which indicates the location of the MSDS information, identifies the OSHA officer for the workplace, and explains how to handle common incidences related to veterinary practice.

4. **Maintain proper secondary container labeling.** You've just received a gallon of 70% alcohol that you use in spray bottles and jars. Any substance or chemical (even water) that you place in any container for ease of use must have a

---

*While you won't catch tick-borne pathogens directly from your patients, you are at risk for exposure to ticks, which transmit these diseases.

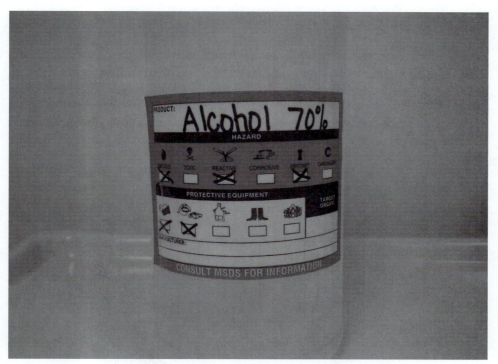

A secondary container label is required any time a chemical is repackaged, such as this alcohol in a spray bottle.
Amy J. Wolff

"secondary container label" that indicates the name of the substance in the new container and any hazard warnings and information specific to the contents.

5. **Provide an emergency plan for response to natural disasters, including evacuation of the patients.** What would you do in the event of a fire, flood, tornado, or other emergency during your workday? All clinics must post a floor plan indicating an evacuation route and a meeting place for the staff should there be a need to leave the building. A plan should be in place to evacuate the patients and continue to provide any care that they may require.

6. **Keep accurate records of workplace injuries and illnesses.** The veterinarian or office manager must document all incidents of injury as they relate to job performance, illness connected to exposure to zoonotic disease, or exposure to substances that are used in the workplace.

7. **Medical waste management.** Along with maintaining a safe work environment are laws that govern the disposal of medical waste. Every day, items are contaminated by blood, pus, feces, and urine. You need to dispose of these in a proper and safe manner. Veterinary practice generates waste material that is potentially dangerous to the public health. You must be careful to dispose of medical waste, including needles, syringes, scalpel blades, bacterial and fungal cultures, discarded vaccine vials, and animal waste from patients undergoing chemotherapy or radiation. Regulations require puncture/leakproof containers for the safe disposal of anything sharp. When possible, you should decontaminate medical waste. If the clinic is unable to do so, you should label it as a biohazard and engage the services of a medical waste management company to provide safe disposal. Although a federal law requires proper disposal of medical waste, there will be differences between how various state and localities comply with this law.

Universal Biohazard Symbol
jezevec10 / Fotolia LLC

In the event that you find a hazard or safety concern, you should immediately communicate your findings to your employer or to the staff member who oversees OSHA compliance in your workplace. Discuss your concerns and ask that somebody investigate the problem. If your employer fails to respond, it is your right under the

Universal Radiation Symbol
Brilt / Fotolia LLC

Universal Laser-in-Use Symbol
T. Michel / Fotolia LLC

OSHA Act of 1970 to request an inspection of any workplace condition that you feel is in violation of the OSHA standards. The law says you cannot be fired, be transferred, or face any employer discrimination for reporting a suspected violation (OSHA, 2013).

Laws Pertaining to the Labor Force

Fair Hiring Practices   As an employee of any business, you are entitled by law to a work environment that complies with fair and legal hiring practices.

Title VII of the Civil Rights Act, 1964   This law, which the Equal Employment Opportunity Commission (EEOC) administers, makes it illegal for employers to discriminate on the basis of race, religion, nationality, or gender. While to date there is no federal law that protects discrimination based on sexual orientation, many states do have legislation in place. The Civil Rights Act also protects you from sexual harassment, making it illegal for employers to make any decision regarding your employment status based on your gender or to use sexual pressure as a condition for your job or advancement. This law also makes it illegal for employers to discriminate by not hiring mothers based on the assumption that they will miss work due to child care issues (EEOC, 2011).

Equal Pay Act of 1963   The Equal Pay Act is administered by the Equal Employment Opportunity Commission (EEOC). The law guarantees that employers must pay equal wages to men and women who perform essentially the same job in the same workplace (EEOC, 2011).

Fair Labor Standards Act of 1938 The Fair Labor Standards Act establishes standards for minimum wages, overtime pay, and record keeping. This law states that "non-exempt" employees will be paid no less than a minimum standard set by the Department of Labor and must be paid at a rate of "time and a half" for work time exceeding 40 hours per week (U.S. DOL, 2009).

You have been getting acquainted with the laws that protect employees in the workplace and assure fair hiring practices. You may not remember the names, the dates, or the federal agencies that enforce these laws, but you should remember that your rights regarding hiring, wages, and nondiscrimination in the workplace are protected.

There are also federal laws in place that protect animals when they are used in research, in exhibition, in competitions, and as companions. Let's take a look at some federal agencies and national organizations that protect animal welfare.

FEDERAL AGENCIES PROTECTING ANIMAL WELFARE

USDA, U.S. Department of Agriculture

The USDA is the branch of the federal government that is dedicated to the management of agriculture and agricultural products. This includes animals used for food and those that are raised for research, exhibition, or sale by breeders.

APHIS, Animal and Plant Health Inspection Service In 1966, the federal government signed the **Animal Welfare Act (AWA)** into law. The Animal Welfare Act regulates the use of animals in research and exhibition, and defines the standards that breeders and dealers must follow for animal husbandry, housing, and transport. The Animal Welfare Act is enforced through the **Animal and Plant Health Inspection Service (APHIS)**. APHIS is a division of the USDA and the agency that is responsible for protecting and improving animal health and the quality of meat, poultry, and egg products. APHIS also employs veterinarians and veterinary technicians to monitor and control animal diseases that are a threat to the nation's food supply.

The AWA has been amended several times to accommodate advances in veterinary medicine and the use of animals in the public domain. These amendments have resulted in laws that demand adequate housing, nutrition, exercise, and veterinary care, as well as laws making it illegal to engage in animal fighting. An example of these laws is the **Horse Protection Act (1970)**.

The Horse Protection Act was enacted in response to the practice of "soring." Soring is the intentional use of chemicals, irritants, or mechanical devices on sensitive parts of the leg and hoof causing the animal to walk in an exaggerated, high-stepping manner. This type of gait is desired in breeds such as the Tennessee Walker, Fox Trotter, and Racking horse breeds. The law makes it illegal to subject any horse to these painful methods.

Another division of the USDA is the **Food Safety and Inspection Service (FSIS)**. The FSIS is responsible for making sure that food-producing animals are not treated with or go through proper withdrawal times for drugs or chemicals that can enter the human food supply. A strict set of guidelines monitoring the use of

The use of irritants and chemicals to produce the high step of the gaited horse has been made illegal by the Horse Protection Act.
Clarence Alford / Fotolia LLC

antibiotics, growth promoters, and other chemicals is maintained under the law to prevent human consumption of these substances.

USDHHS, U.S. Department of Health and Human Services

The Health Research Extension Act of 1985, under the Office of Laboratory Animal Welfare, establishes guidelines for the care of animals in biomedical and behavioral research. Any investigators applying for a federal grant to conduct research involving animals must submit documents to describe how they will use the animals and what type of veterinary care the animals will receive. The National Institutes for Health (NIH) developed the guidelines for housing, veterinary care, nursing, pain control, anesthetic procedures, and the use and compliance of tranquilizers and other drugs (NIH, 2004).

NATIONAL ORGANIZATIONS PROTECTING ANIMAL WELFARE

National organizations cannot pass laws, but they may advocate or lobby the legislature for laws to be enacted. These organizations have elected governing bodies but offer the public the chance to offer support through donations and membership. There are many different organizations, all with their own particular focus and area of interest in animal welfare. An example that may be familiar to you is the Humane Society of the United States.

Humane Society of the United States

The **Humane Society of the United States (HSUS)** is the nation's largest animal protection organization. Created in 1954, its mission includes supporting legislation by advocating for better laws to protect animals. The HSUS conducts campaigns to reform animal industries, end animal fighting, and protect wildlife; its emphasis is on education, and legislation, as well as rescue and shelter for companion, research, and farm animals, horses, and wildlife (Humane Society, 2011).

American Society for the Prevention of Cruelty to Animals

The **American Society for the Prevention of Cruelty to Animals (ASPCA)** is another animal welfare group similar in focus to the HSUS. The ASPCA was the first organization to address the humane treatment of animals and actively work to prevent cruelty and abuse. The ASPCA sponsors educational events, supports rescue and shelter, and works to pass legislation that protects animal welfare (ASPCA, 2011).

PROFESSIONAL ORGANIZATIONS

Professional associations offer veterinarians and veterinary technicians an opportunity to share their common interests in specific areas. While the main focus of these associations is continuing education in disease recognition and treatment, they are the voice of the veterinary profession for recognizing and raising awareness about the needs and issues of animals.

Organization	Focus
American Veterinary Medical Association	AVMA website A resource for animal health topics, research, animal advocacy and legislation, information for pet owners and related topics. Professional development and career center tools are available
National Association of Veterinary Technicians in America	NAVTA website Representing and promoting the profession of veterinary technology
American Association of Equine Practitioners	AAEP website Equine health topics for veterinarians, students and horse owners. Equine welfare, legislation, sports medicine and event calendars are among the Associations many resources
American Association of Bovine Practitioners	AABP website Dedicated to improving the well being of cattle, and the economic success of cattle producers and the cattle industry
Association of Avian Veterinarians	AAV website Promotes the advancement of avian medicine and responsible care of birds.
American Association of Feline Practitioners	AAFP website Dedicated to improving the health and welfare of cats through scientific research and high practice standards.
American Association of Laboratory Animal Science	AALAS website Promotes the responsible care and use of laboratory animals

These are a few examples of the governmental agencies, organizations, societies, memberships, shelters, and rescue groups that are actively engaged in animal welfare in food production, research, and private ownership. There are numerous bills and pieces of pending legislation on many "hot" animal topics, including puppy mills, factory farming, and the fur industry. Explore some of this legislation. Discover the scope of issues, and familiarize yourself with the many organizations representing animal welfare (Animal Law, 2011).

COMMON LAW AFFECTING THE PRACTICE OF VETERINARY MEDICINE

Malpractice and Negligence

Malpractice and negligence are examples of common law that affect veterinary practice. The terms *malpractice* and *negligence* are almost the same. The terms apply when you have failed to follow an accepted standard of care and as a result the patient is injured. When this happens, people often **sue** to solve a dispute or seek compensation. Filing a lawsuit is commonplace in our society, and the practice of veterinary medicine is no exception. A client can claim malpractice or negligence against the veterinarian or the veterinary technician if he or she can prove that a valid Veterinary-Client-Patient-Relationship existed, that the "standard of care" was *not* met, and that injury occurred because the level of care was inadequate. By law, once the veterinarian has agreed to care for a pet, he or she is obligated to meet a standard of care that a person of similar training under the same circumstances would provide. The burden of proof lies with the person bringing the lawsuit, but a charge of malpractice or negligence may be difficult to defend if the veterinarian and staff did not maintain accurate documentation. It is in everyone's best interest, most importantly the patient's, to always have a clear, concise, accurate, and up-to-date medical record and consistent client communication to avoid these situations.

You have just finished reading about laws and regulations that affect our actions and obligations to veterinary patients, our clients, and the public. Legislative laws are clear. We know what is legal and illegal.

VETERINARY ETHICS

As we begin our discussion of ethical behavior, let's define the concept of *ethics*. When someone is thinking about "ethics" or "morals," they are usually considering the action that he or she feels is the "right" or "proper" thing to do. Your personal values determine ethical principles, which usually have a strong foundation in the "rights" and "wrongs" that your parents taught you as a child. Religious, cultural, and personal experience help form your ethical values. It is important for you to understand that people of similar backgrounds and cultures can have very different ethics. This is an important concept for you as a veterinary professional, because your coworkers, clients, and supervisors may disagree as to what defines the ethical treatment of animals.

There are numerous issues within the veterinary profession that test our personal ethics. Often, it is a challenge to see the reasoning and understand the feelings of people who have opposing viewpoints to your own. It is in the best interest of the practice to accept that your clients and the people you serve in the community will not always see things your way.

Personal Ethics: The Role of Animals in Society

The concern for the well-being of animals and the idea of what is right and wrong opens the door to an important topic. Ethical dilemmas in veterinary practice often arise because animals are *legally* defined as property.

Let's discuss the concepts of *animal welfare, animal rights, ownership,* and *guardianship.*

Animal Welfare   The concept of animal welfare acknowledges that we will use animals for food, fur, and research, but advocates that all animals will have appropriate care, suffer minimal pain, and will have a fast, humane death.

Animal Rights   This concept says that animals have the same rights and, therefore, are essentially equal in consideration to humans. Supporters of this concept believe that we should not eat animals or use them for any type of research. We should also not hunt, trap, or use animals for sport.

Animal Ownership   The law defines animals as the *property* of the owner. Provided that the owner is not negligent or cruel and that he or she obeys local ordinances and meets the basic needs of food, water, and shelter, the owner can do what he or she wishes with his or her animals.

Animal Guardianship   Some people have a preference for the term *guardian* as it relates to their relationship with their animals. They feel it is a better expression of the deep bond that they feel with their pet. The term *guardian* seems a softer word, with deeper implications of the emotional responsibility than the term *owner*. However, the term guardian has very different legal implications. If a person is an animal's guardian, the term implies that the animal's needs come before those of the owner. Health situations may arise where an owner may choose to treat a pet's disease or condition when in fact the condition may, unfortunately, warrant euthanasia. For example, a longtime pet owner brings his beloved golden retriever to the veterinary hospital. The veterinarian diagnoses congestive heart failure. The prognosis is poor. The "owner" would like to consider euthanasia to avoid the pet suffering from a deteriorating condition. The "guardian" would have to hospitalize and treat the pet even though he or she might be financially or emotionally opposed to treating a pet with a life-threatening disease (AVMA, 2012).

AN EXERCISE IN ETHICS: THE FARM DOG VERSUS THE CITY DOG

In city and suburban areas, a common expectation is that a dog is a house pet, primarily to provide companionship. The majority of these pets live indoors, and owners often treat them as part of the family. The animals are dependent on their owners to provide food, play time, exercise, and mental stimulation. Most of these animals are spayed or neutered, and local laws or other regulations may require their confinement to a house and yard. Socializing with other dogs occurs in dog parks, doggie day care, or other communal areas. To comply with local laws, owners walk their dogs on leashes and must clean up their waste. Often, these companions attend obedience classes so that they can become good canine citizens and learn to behave properly around people and other dogs. Owners often buy collars, coats, chew toys, and personalized items for these pets.

In rural areas, the family dogs might lead a vastly different life. Bed might be a barn or a dog pen. During the day, they might roam freely. They often interact sexually and reproduce at an early age. Grooming and training may be less of a concern for the owner. They may use dogs for specific jobs, which include hunting, guarding, herding, and protection.

Susan Schmitz / Shutterstock

These two dogs live very differently based on their owners' lifestyles.
beerfan / Fotolia LLC

If we talk to the owners of these dogs, the suburban dog owner might tell us that keeping a dog outside is unethical, because it has to put up with harsh weather conditions, is isolated from the family, and suffers psychologically. Furthermore, the animal has no supervision, which increases the risk of injury, property damage, and unwanted breeding.

The rural dog owner might say that the suburban dog does not exercise properly or have mental stimulation, and that they never have the chance to just "act like dogs." These views differ because they are based on what's normal for the culture of the area.

Imagine how a little Poodle or Shih Tzu fresh from the groomer, with a scarf and painted nails, would look running around the farm pond. Try to imagine a dog that's been out roaming all day, rolling in dirt, and snacking on garbage would smell sleeping in bed with you. It would seem that each owner has his or her own viewpoint. What do you think? Remember, there's no right or wrong, only your own personal views.

LEGAL VERSUS ETHICAL

Let's examine some common scenarios in veterinary hospitals to help you recognize the difference between the two.

Scenario 1: Imagine that you have completed the first phase of your veterinary education. You are a brand-new graduate and have landed your dream job at All Pets Veterinary Hospital. During your first week at work, Mr. Harvey comes into the practice with a sick puppy. He is a new client and has walked in without an appointment. You haven't seen many sick animals yet, but you know enough to recognize the clinical signs and smells of Parvovirus. This is one sick little dog, and he is going to need treatment—fast.

Your veterinarian is busy examining the patient and educating the client. She has asked the veterinary assistant to prepare an estimate for hospitalization and supportive care. The estimate exceeds Mr. Harvey's financial expectation. He is not prepared to spend as much money as predicted by the estimate or to leave half as a deposit. Everyone knows that this puppy's chances of survival are small without supportive care. The client is angry and accuses the veterinarian and staff of not caring about animals. He says that their only concern seems to be money. He picks up the puppy and leaves. The veterinary team members are quiet as they anticipate the fate of this puppy.

Sue McDonald / Shutterstock

Scenario 2: Mrs. Bradley is a good client of All Pets Animal Hospital. She has taken good care of Scrappy, her six-year-old terrier mix, and has always provided necessary health care. This week, Mrs. Bradley decided to adopt another dog from a local shelter to keep Scrappy company. A week or so after Scruffy joins the family, he develops a cough and diarrhea. Mrs. Bradley calls the hospital asking for medication. You advise Mrs. Bradley that the hospital cannot provide medication for a patient it has never seen. Mrs. Bradley does not take well to this news. She reminds you that she has been coming to your veterinary hospital since it opened. and now she has tried to do a good thing by providing a home for a dog that would have otherwise been put to sleep. She's short on money this week after the adoption fee and supplies and needs All Pets Hospital to do their part for Scruffy.

Scruffy, Mrs. Bradley's new dog.
biglama / Fotolia LLC

In both of these scenarios, clients have come to or contacted All Pets Animal Hospital seeking treatment for their pets, but each situation represents very different circumstances. Let's review them individually.

In scenario 1, Mr. Harvey has come to the hospital with a sick puppy. He would like the veterinarian to treat the puppy, but the estimate for care has exceeded what he was expecting to pay. He feels that the veterinary hospital has a moral and ethical responsibility to treat the animal whether he is able to cover the cost or not. In this situation, the hospital has no legal obligation to treat the puppy, but Mr. Harvey feel it is the "right" thing to do.

In scenario 2, Mrs. Bradley has surely expressed her feelings. She expects All Pets to contribute to her efforts to adopt a stray dog by providing free medication. However, in this scenario the veterinary hospital would be breaking the law in doing so. It is illegal for a veterinarian to prescribe medication for any animal he or she has not seen. Dispensing medication to Mrs. Bradley's new pet is a legal issue.

Can you think of someone whose ethics regarding animal welfare differ from your own? You might stop to consider some of the many debates that are ongoing in our profession. Many of these issues are very controversial and can touch a deep emotional core.

ETHICAL AND CONTROVERSIAL ISSUES IN VETERINARY MEDICINE

As you read over the descriptions about these issues, ask yourself, "What do I think about these topics? If I had to take a stand, what side would I take?" There is no right or wrong answer; there's only your feeling of what is right or wrong from your point of view.

1. **Animal research.** Is it ethical to use animals to test pharmaceuticals, cosmetics, or medical procedures that may help both human and veterinary medicine? Is it OK to use rats and mice, but wrong to use cats, dogs, or primates? Find some information about types of research using animals, and see if you can support it or disagree with it. Do you know someone who has benefitted from research that used animals?

2. **Raising animals for food/fur.** Many people are turning away from a lifestyle that includes eating or wearing any products that are derived from animals. How do you feel about consuming meat, milk, or eggs? What about wearing clothing made from leather or fur? Is it ethical to raise animals to provide these products to consumers?

3. **Wildlife habitat preservation.** As populations expand, people often have to tear down natural habitats to make room for cities, roads, and other types of developments. This loss of habitat threatens the survival of many animals. Is it better to preserve natural environments and habitat, or is it better to allow people to use these resources to ensure that they have homes and can continue to make a living?

4. **Veterinary cosmetic surgery.** Certain breeds of dogs that have a characteristic look because their owners have elected to crop their ears or tails. Examples of these breeds are Great Danes, Doberman pinchers, Schnauzers, Yorkshire terriers, Rottweilers, and Boxers. Is it right to crop ears and tails to modify a dog's appearance? What about declawing cats? Some clients will tell you that they cannot have a cat unless it is declawed. Others will say it's wrong to deprive a cat of its natural defenses.

What are your feelings about raising animals solely for the use of their fur?
mythja / Shutterstock

5. **Euthanasia.** Do we, as veterinary professionals, have the right to end a life? Some say euthanasia is a blessing to end pain and suffering and provide a swift, painless death. Others will argue that it is not up to people to determine when life is to end.

6. **Free medical treatment for clients who can't afford care.** Let's return to the first scenario above. You had a client at All Pets Veterinary Hospital,

The destruction of native habitats puts animal populations at risk.
Dr. Morley Read / Shutterstock

Should animals have elective cosmetic surgery to alter their appearance?
ELPSTOCK / Fotolia LLC

Mr. Harvey, who presented a puppy that was in need of veterinary care. The client could not afford the treatment plan. Is it ethical for veterinary professionals to deny care to patients who cannot afford treatment? This is a difficult question, and you must give it considerable thought. Providing free medical care can quickly drain the hospital's resources. Should All Pets sacrifice the ability of the hospital to be open for other patients? Can the hospital risk not ordering supplies that it cannot pay for due to lost income? Do you want to take a pay cut? Do you want to see the parvo puppy leave without proper care?

Are you beginning to see the arguments on both sides of these ethical questions? Sometimes it is as if you could go either way. It is up to you to decide what is important to you and how you will adhere to your ethical principles.

THE FIVE FREEDOMS

In 1965 a British report authored by Roger Brambell on the use of animals in agriculture served as the basis to create a set of guidelines now known as "the five freedoms." These freedoms are not laws, but principles to apply to the care of all animals under human supervision. Many organizations have adopted these principles as the ethical core of our relationships with animals.

- Freedom from thirst and hunger
- Freedom from discomfort
- Freedom from pain, injury, and disease
- Freedom to express normal behavior
- Freedom from fear and distress

PROFESSIONAL ETHICS

Although you may not know your classmates or other veterinary professionals well, you share an understanding of the deep bond that you feel for animals and the instinct to protect their welfare. This compassion is the foundation for the expectations of your professional behavior. Veterinary professionals are held to very high standards, both by society and each other. Collectively, we want to have the trust and confidence of our clients and the public. Your clients will expect that you will take care of their "babies" as you would your own. You are responsible for public health. Because you have accepted that role, you should strive to advance your skills, training, education, and image as a professional. The National Association for Veterinary Technicians in America (NAVTA) has published a list of nine ethical principles on their website that are the basis for your ethical and professional conduct.

Model your behavior and attitudes from these principles. They will serve as a framework for your entire career.

SUMMARY

As with all professions, veterinary medicine is regulated by federal, state, and local laws. As part of the health care team, you must familiarize yourself with these rules so that you can give proper care to the patient, keep the client informed, and help the hospital comply with the law.

There are numerous regulations that protect you in the workplace. The goal is to minimize hazards and risks by providing you with the training, resources, and equipment to accomplish this goal. You are protected against discrimination for age, race, nationality, and gender when applying for a job.

The government also protects animals. There are laws in place to assure that owners house and feed them properly, that they receive adequate veterinary care, and that the owners meet the animals' psychological needs. Laws also vary by state and locality. Regional and cultural differences account for these variations.

Your career will introduce you to many people who have different ideas about the proper care and use of animals. Veterinary technicians must show tolerance and understanding. This is not a compromise of your own personal values, just a mature acceptance that people have different opinions. Change begins with education, and that is where you can make a world of difference.

TEST YOUR KNOWLEDGE

1. Mrs. Simmons comes to the clinic to have her puppy's ears cropped. She likes the way Great Danes look with their erect ears and alert appearance. Your clinic does not provide this service. Is this a legal decision, an ethical decision, or both? Hint: Check your state practice act for the practice statutes where you live.

2. You're late on Monday morning, and you are speeding to work when you are pulled over by law enforcement. As you pull off the road, you run over a bike laying half in and half out of the street. You are given a ticket for speeding, and your neighbor is claiming that you owe him for the damage to the bike. You say the bike was too far into the street to avoid. Which of these two problems is a case of legislative law, and which is a case of common law? Explain your answer.

3. Your boss frequently asks you to take radiographs with another technician, and there is only one set of gloves available, so the two of you share them.

You asked your employer to buy another set of gloves, but that was more than a month ago. What organization enforces safety by setting standards for employers to minimize hazards in the workplace? What are your options if your boss refuses to buy another set of gloves?

4. What is the name of the federal law that protects animals that are used in research, shows and competitions, biomedical studies, and transportation? What types of standards for animal welfare are set by this law?

5. To prepare yourself for the challenges of veterinary practice, you will need to develop some healthy strategies for dealing with outcomes of ethical situations that conflict with your views. What types of activities do you feel will help you cope with situations that are contrary to your personal values?

6. Choose one of the controversial topics in veterinary medicine and write a paragraph describing your feelings. State whether the topic is a legal issue or an ethical one.

BIBLIOGRAPHY

Animal Law. 2011. "Home Page." Accessed May 6, 2012 from the Animal Law: International Institute for Animal Law website.

ASPCA. 2011. "Home Page" Accessed May 7, 2012 from the America Society for the Prevention of Cruelty to Animals website.

AVMA. 2012. "State Advocacy Issue." Accessed June 1, 2012 from the American Veterinary Medical Foundation website.

EEOC. 2011. ""Equal Employment Opportunity Commission." Accessed June 1, 2012, from the U.S. Equal Employment Opportunity Employment Commission website.

NAVTA. 2009. "A Model Practice Act." Accessed June 1, 2013 from the NAVTA website.

NIH. 2004. "Office of Laboratory Animal Welfare." Accessed November 1, 2011 from the National Institute of Health website.

Humane Society. 2011. "About Us." Accessed May 7, 2012 from the Humane Society of the United States website.

OSHA. 2012. "Regulations" Accessed June 4, 2013 from the Occupational Health and Safety Administration website.

Thorpe, W.H. 1967. Brambell Committee Report, Command Paper 2896H.MSO, London (1965), pp. 71–79.

U.S. DOL. 2009. "Overtime Pay." Accessed May 22, 2012 from the United States Department of Labor website.

4

The Animal Industry

The author gratefully aknowledges the Chapter 4 contributions of Michael Richards, DVM.

Learning Objectives

At the end of the chapter, you should be able to:

- Explain and give examples of the economic significance of the animal industry.
- List and describe the benefits of companion animal ownership.
- Identify business trends in the companion animal industry.
- Outline the advice that you would give a client in selecting the appropriate pet.
- Recognize and give examples of irresponsible breeding practices.
- Discuss the economic basis for livestock production.
- Summarize technological advances in biomedical research.

INTRODUCTION

Students of veterinary technology often express a deep regard and emotional attachment to animals and the enrichment that they bring to their lives. You may have similar feelings. You would feel incomplete without the close presence and daily interaction with animals. You must now become familiar with the business that is centered on animal ownership. In Chapter 2, you learned about the principles of running a veterinary hospital as a business. Along with veterinary care, there is an enormous industry with significant impact on our economy. Businesses and activities that generate profit from the use of animals (including animal supplies and services), the sale of animals, and the sale of animal products are all parts of what we call the **animal industry**. When a veterinarian charges for medical services, a pet store sells dog food or bird toys, a farmer raises cattle, or a drug company sponsors a canine agility competition, they are participants in the enterprise that makes up the animal industry. Veterinary medicine, pet product manufacturing and distribution, animal feed production, commercial farming operations, and animal breeding are just a few examples of the various facets of the animal industry. For the purpose of this chapter,

yadamons / Fotolia LLC

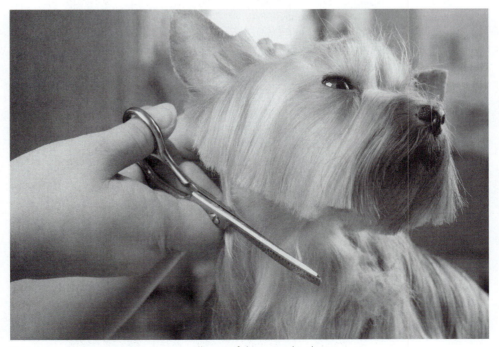

Farming and grooming services are all part of the animal industry.
Scorpp / Shutterstock

Veterinary services are also part of the animal industry.
melis / Shutterstock

we will be examining three components of this field of business. They encompass the topics most relevant to your career as a veterinary technician.

- Companion animals including the exotic animal trade
- Livestock
- Animal used in research

Before we examine these components, let's take a look at the roots of the "business of animals."

THE ANIMAL INDUSTRY: ANCIENT BEGINNINGS

Since early human history, animals have provided us with food, labor, protection, companionship, and even entertainment. The animal industry may well have had its beginnings thousands of years ago when the first ancient humans hunted animals and then traded furs with neighboring tribes for other necessities. Evidence indicates that dogs were domesticated from gray wolves for protection, an ancient "security system" to alert early tribes of intruders. *Domestication* means that an animal has adapted to human living conditions and that humans have chosen them for breeding, to pass on desirable traits such as herding ability or milk production.

chippix / Shutterstock

Fotolia XXIV / Fotolia LLC

This relationship was mutually beneficial for the humans and the wolves: the animals were afforded shelter and food, and in return provided protection and companionship. As human communities became more agricultural and less nomadic, needs changed. These communities then domesticated other types of animals to help lessen the burden of farming. Our ancestors domesticated the animal that we now know as a cow from several wild species—for milk, meat, and labor. Dogs became our aids in herding instead of hunting.

Human activity based around animals was now geared more toward raising animals for food and the sale or trade of farm products. Our basic needs have not changed much over time, but we have expanded the scope of the role that animals play in our lives and the number and type of animals that we include in that role.

COMPANION ANIMALS

What Is a Companion Animal?

It is easy to think of a companion animal as a pet that we keep for company and amusement. The definition is really a bit more complex. Companion animals offer many people emotional and psychological support. They may be a critical component in the life of the owner, especially if the pet fulfills a service role. Most, but not all companion animals, return the owner's affection and stay faithful to the household. If an owner invests time and effort in a living being's care and welfare, there is an emotional reward that results in a bond. In your career in the veterinary clinic, you will be surprised more than once at the types of pets people choose for company.

Exotic Animals

Veterinary medicine broadly uses the term *exotic* to describe any animal that is not a dog, cat, horse, or cow. This leaves the classification of *exotic animal* to include many species. The profession groups small mammals, big cats, birds, reptiles, fish, and insects together in the category of exotic animals. It is becoming more common for people to keep exotic pets, and this opens up many heated debates and controversies about the types of animals that are suited to captivity and domestication.

lolloj / Shutterstock

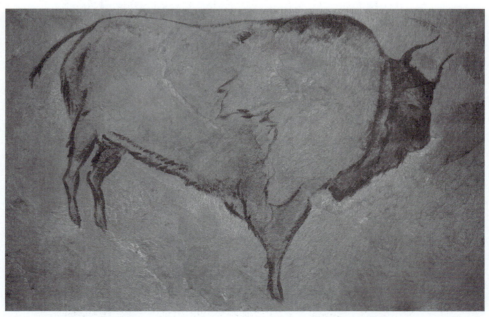

Ancient art shows the relationships between early humans and animals depicted in great detail.
Awe Inspiring Images / Shutterstock

This dog has been domesticated and selected for its herding ability.
Arkadiusz Komski / Shutterstock

Serg Zastavkin / Shutterstock

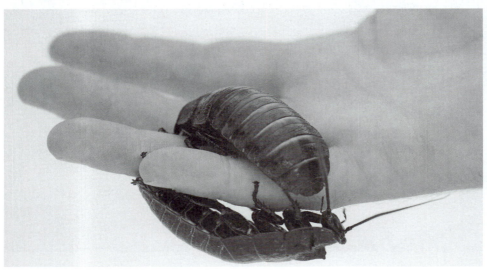

Dhoxax / Fotolia LLC

Kamil Cwiklewski / Fotolia LLC

MaZiKab / Fotolia LLC

People choose all types of animals as pets.
Eric Isselée / Fotolia LLC

Economic Aspect

The business that supports animal ownership has become a significant part of the U.S. economy. Take a closer look at the amount of money that people spend on companion animals so that you have an idea where the dollars are going. The figures below are from the American Pet Products Association.

ESTIMATED 2013 SALES WITHIN THE U.S. MARKET

- According to the 2011–2012 APPA National Pet Owners Survey, 62% of U.S. households own a pet, which equates to 72.9 millions homes.
- In 1988, the first year the survey was conducted, 56% of U.S. households owned a pet as compared to 62% in 2008.

For 2013, the association estimated that U.S. pet owners would spend $55.53 billion on their pets.

Estimated Breakdown:

Food	$21.26 billion
Supplies/OTC medicine	$13.21 billion
Vet care	$14.21 billion
Live animal purchases	$2.31 billion
Pet services: grooming & boarding	$4.54 billion

For the time period 2001–2011, spending on pets and pet services has doubled from $28 to $52 billion dollars per year.

APPA, 2013b. "Industry Trends" Accessed July 28, 2013 from the American Pet Products Association website.

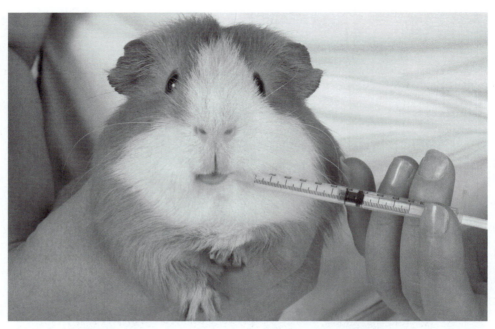

mariesacha / Fotolia LLC

The figure that reports the amount of money that U.S. pet owners spend on veterinary care should be of interest to you. In 2013, the estimate is 14.21 billion. In other words, pet owners spend almost one-third of the total amount for veterinary care (APPA, 2013).

Notice that supplies, over-the-counter medicines (OTC), and pet services make up a slightly larger percentage than does veterinary care (32%). This is a direct reflection on how the status of the family pet has changed. Previously, pets lived outside and had to find their own shelter. Food might have consisted of leftover scraps from the house and whatever they could scavenge or hunt. Rarely would an array of toys, designer collars, day camps, spas, or pet hotels be in the family budget. How does this compare to the life that your pet leads? Many animals share the home equally with their owners, with unlimited access to the house, and they always have an abundant supply of food and water. What do you think has caused this shift in the lifestyle of the average pet?

Imagine for a moment that you have just purchased a new puppy from a local dog breeder. Within the first month of owning your new dog, you would likely make a visit to your veterinarian for vaccinations, flea and heartworm prevention, diagnostic tests such as a fecal exam, and, later, neutering. You would certainly go to the pet store to buy dog food, a leash and collar, and some toys for your new puppy. Perhaps, after he is fully vaccinated, you would take him to the groomer for a bath.

In the space of one month, you have utilized at least six different participants in the companion animal industry. Dog breeders, veterinarians, veterinary diagnostic laboratories, drug manufacturers, and pet stores and groomers have all played a part in the adoption of your puppy. As you become a member of a veterinary team, you will have a role in recommending to your clients products and services that you believe will enhance pet health care and responsible pet ownership. Representatives from pharmaceutical companies, trainers, groomers, and private entrepreneurs will visit your workplace to introduce your hospital to the products and services they provide. Each is trying to fill or compete for a niche in your hospital. You will find products that you love to use because they make your job easier or they are the most effective. By selecting the products that you feel meet your patient's needs and recommending them to your clients, you are supporting the animal industry.

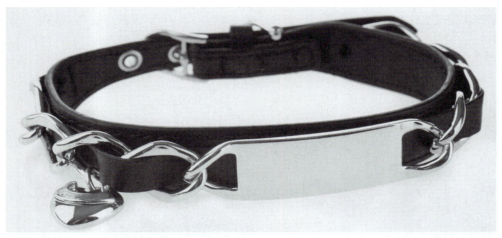

Dean Bertoncelj / Shutterstock

Designer luxury collars are a must-have for some dog owners.
Tooties / Shutterstock

Benefits of Animal Ownership

While you concentrate on promoting and maintaining animal health, you should not lose sight of the fact that human health improves from animal's companionship. The Center for Disease Control reports that pet ownership has the following benefits: Pets can *decrease* your:

Blood pressure

Cholesterol levels

Triglyceride levels

Feelings of loneliness

Pets can *increase* your:

Opportunities for exercise and outdoor activities

Opportunities for socialization (CDC, 2013)

Congratulations on your new puppy!
Kirill Kedrinski / Fotolia LLC

Puppy's first vet visit.
gorillaimages / Shutterstock

Appropriate chew toys are a necessity.
Mila Atkovska / Shutterstock

Pet owning has many health and social benefits.
Mona Makela / Shutterstock

The current boom in pet ownership, which has resulted in $50 billion in sales, is positive in many aspects. Your job is one of them. Promoting and protecting animal health and educating the public has never been more important. It is essential that you be able to advise clients about the best way to care for their animals' physical and psychological health and how to obtain them from reputable sources.

Client Education

Most people base a pet purchase on impulse or emotion. The best situation occurs when your client comes for advice *before* they bring home a pet, but most of the time your client already will have purchased a new puppy, kitten, or bird before they come to the hospital. People will often do *more* research on buying a car or appliance than they do for a pet. What advice can you offer a client who comes to you seeking information about selecting an appropriate pet from a legitimate source? Here are some key points that you can discuss with your client to help him or her make the best possible decision.

1. Do your research! Pick an animal that suits your lifestyle and energy level, your capabilities, and your finances. There will be costs involved in maintaining the health of the animal besides initial purchase. Almost every animal species has breeds or types that are *high maintenance* and may require a larger commitment to care and husbandry.

2. Is the pet prone to any specific medical conditions? It is important to educate your clients in the diseases that are common to the pet they intend to purchase. This allows them to question the breeder about any efforts they have made to eliminate the trait from the breeding stock. For example, German shepherd dogs have a high incidence of hip dysplasia. Maine Coon cats often suffer from heart disease. Great Danes have a higher incidence of gastric dilatation (bloat). Future pet owners should know about prevalent conditions so that they can be vigilant in their search for information regarding diseases affecting the welfare of their prospective pet.

The ambilobe panther chameleon will need a controlled environment and specialized diet.
Cathy Keifer / Shutterstock

3. Is it legal to own the animal in the community? Some localities have breed bans or require specialized containment such as eight-foot fences or enclosed kennels. Are there specific covenants or restrictions in the neighborhood? Does living in rental property exclude pet owning as an option? If not, will there be additional fees?

4. Do you have children? Will the animal be appropriate for their age range?

5. Are there individuals in the household who have allergies or sensitivities to pets? Will this be a major obstacle in the care and housing of the pet?

6. Can you obtain the pet from a reputable source? Armed with proper information about any conditions that the pet may develop, ask the breeder/supplier about any measures they have taken to limit the chances of inheriting a genetic problem. For companion animals, inquire how the breeder screened the parents before allowing them to produce a litter. Did the breeder have a radiologist examine and evaluate the dog's hips for signs of hip dysplasia? Did a veterinarian screen the eyes for heritable eye conditions? Did the breeder get a temperament test for the animal? Did the horse have a lameness exam? If the answer to any of these questions is "No, we've never had any problems before . . ." please exercise caution.

7. Does the breeder question your ability to care for an animal? Does he or she make you sign a contract? Can you return the animal if you cannot care for it or it does not pass a veterinarian's health evaluation? Legitimate breeders want to know where their animals are going. Many want the owners to return them if circumstances change. Breeders may require the owner to spay or neuter the pet. Shelters and rescues are likely to do the same.

8. Are the parents available for you to meet? This will not always be the case, but if it is possible, take a look at the sire and dam to assess them for general conformation, temperament, and socialization. Can you see the animals' living conditions? Are the animals clean and well cared for or living in poor conditions? Can you examine copies of the veterinary records, especially those that indicate that screening exams have taken place?

9. If possible, leave children at home when selecting a pet; children make their choices based solely on emotion. There has to be a component of rational thought. Ideally, select the animal that you feel is most suitable and then introduce children to the potential new family member.

10. Take extra care when buying pets from pet stores. Some of these businesses are reputable and take measures to ensure their animals' health and well being, while others are essentially a clearinghouse of animals that come from production farms. Carefully evaluate any "health guarantees"; it is very difficult to guarantee the health of a living organism.

Sources of Animals

Let's examine a few issues so you begin to develop a basic understanding of the animal industry as it relates to the sale and care of animals. Dogs, cats, horses and cattle, and exotic animals must have a source from which you can purchase them. If you want a green iguana, you will not likely find vendors giving them away like puppies at your local gas station. Where do you obtain a green iguana—or a toucan or an Emperor angelfish? Most people begin their search by accessing the Internet to look for breeders and other sources.

Unfortunately, there are many unregulated, irresponsible, and illegal factors in the companion animal industry. An example of this is the puppy mill. "Puppy mills" exist for the purpose of breeding large volumes of dogs primarily for sale to pet stores

Conditions in puppy mills are often substandard and threaten the physical and psychological health of the animals.
KROMKRATHOG / Shutterstock

and through the Internet. Dogs from these facilities are often poor representations of their breeds or have multiple health problems. Frequently, the breeders keep the animals in strict confinement and provide them with minimal care or social interaction. Poorly socialized animals usually pass these characteristics to the offspring, making the puppies fearful or aggressive at a young age. In order to maximize profits, breeders often expect females to produce two litters a year until well into their senior adulthood. Breeders cut costs in a number of ways: overcrowding, poor-quality food, and inadequate veterinary care are common. Despite the poor conditions and lack of ethical standards, puppy mills continue to thrive. A lack of information and education for the consumer is one of the main reasons that these businesses remain profitable. Sometimes people may simply not be aware of the substandard care that these animals receive; in other cases, they *are* aware, but often buy a puppy out of compassion to remove it from the poor environment. The puppy mill industry is rooted in deceiving the buying public into thinking that they are purchasing a quality, purebred dog.

One of the concerns in the exotic animal trade is how people obtain the animals. Frequently, if the businessperson cannot domestically breed the animals, he or she captures them from the wild, often at a young age, and ships them to brokers or pet stores for sale to customers. There can be a high mortality rate from the stress of capture and handling, and a poor success rate in getting the animal to adapt to a life of confinement.

What factors do you think influence the consumer to buy an exotic pet? Commonly, it is the excitement of owning a pet that is different and unique. Exotic pets are often small and animated and do not require much space. Or they may be large, stunning, and intimidating. Some, like snakes, do not require much daily care and do not require exercise. While there is no lack of motivating factors for purchasing an exotic pet, there can be a serious lack of understanding of appropriate care for the animal. People rarely take time to research the needs of an exotic animal, even the ones

Saltwater fish are often removed directly from their native habitats and shipped to pet stores.
cbpix / Shutterstock

that you see commonly. For example, people frequently buy hamsters as the first pet for children because of their cute appearance and small size. However, hamsters are nocturnal and can be very disagreeable about engaging in play during the day when a child might want company. They can be ill-tempered and bite. A better choice might be a rat. They are intelligent and usually quite agreeable to human companionship, but people often dismiss them too soon as a pet because of the appearance of their tail and their persistently yellow teeth.

There is often a large "knowledge gap" about how to appropriately house, feed, and care for an exotic animal. Nutritional deficiencies, improper environment, and poor handling result in the illness and death of many exotics. This scene is played out many times in veterinary hospitals when owners present iguanas for bone disease because the owners have fed them a diet consisting mostly of lettuce, or rabbits with intestinal disease because of a diet consisting of pelleted feed.

Impulse purchases of exotic animals can lead to serious consequences. Consumers should make sure the pet is legal to own, whether ownership will affect their homeowner's insurance, and if the pet is suitable to be around children. You must also consider the animal's life span. Some birds and reptiles can live well into their fifties and sixties, requiring some type of arrangement for continuing care if the pet outlives the owner.

Public Health and Safety

As you have seen and heard many times from news stories, not all animals are well suited to living with humans, and tragedy occurs. Primates such as chimpanzees and large cats such a lions and tigers are rarely safe pets to keep in a household. While they may be endearing as youngsters, they mature into dangerous animals driven by reproductive instincts, prey drive, and territoriality. The baby alligator that you purchase at the county fair that you can keep in an aquarium is quite a different pet than a six-foot mature male gator that is bigger than your bathtub.

Despite their adorable appearance, hamsters are often ill-tempered.
Dimco / Fotolia LLC

In addition to the injury that exotic animals can cause, there is also the concern of the potential for *zoonotic* disease, one that can pass to humans from animals. Animals coming from random sources, exotic animal auctions, or irresponsible breeders can cause serious illness. Every year, there are approximately 70,000 cases of Salmonella reported from contact with pet reptiles (CDC, 2012).

In 2003, a group of prairie dogs shipped to the United States from Africa caused an outbreak of monkeypox (CDC, 2012).

Despite these concerns, exotic animal ownership is growing, although it is hard to estimate accurately the number of these pets in households. Many owners obtain

anatolypareev / Shutterstock

animals through nontraditional sources or illegally. In 2013, the American Pet Product Association reported the following statistics for the total number of pets owned in the United States (in millions):

Dog	83.3
Cat	95.6
Bird	20.6
Horse	8.3

Source: APPA's 2012–2013 National Pet Owners Survey (APPA, 2013)

This Yellow-footed tortoise can live up to 100 years.
Ryan M. Bolton / Shutterstock

Heiko Kiera / Fotolia LLC

Juvenile exotics may quickly grow into dangerous pets that exceed the owners' resources.
© Eric Isselée/Fotolia

Because of the easy accessibility, unique appeal, and inconsistent laws regulating the ownership of exotic animals, these figures are likely to increase.

We should not end our discussion of companion animals without acknowledging the efforts of responsible pet breeders who are devoted to the health and welfare of their animals. There are many people who follow sound breeding principles, are licensed by the United of States Department Agriculture (USDA), work in concert with their veterinarians, and carefully consider the consequences of producing a litter of puppies or kittens, a foal, or a clutch of snake eggs. The truly responsible breeder always has the animals' best interests before his or her own.

LIVESTOCK INDUSTRY

When you hear the term *livestock*, it refers to any animal raised in an agricultural setting for the purpose of providing food, fiber, labor, or other resources. In the United States, the term typically refers to cattle, horses, pigs, sheep, goats, rabbits, and poultry. Producers can raise less traditional species as well. Bees, worms, and fish are also marketable commodities. (There are various opinions about whether or not to include horses in the term *livestock*. In some areas, cultural trends and attitudes about horse ownership have placed horses in the category of companion animals.)

branislavpudar / Shutterstock

Dhoxax / Shutterstock

Chickens, sheep, and fish are commonly raised in agricultural settings and contribute significantly to the United States economy.
Vladislav Gajic / Shutterstock

Raising livestock is a commercial enterprise. This means that the primary intention of raising these animals is to derive a profit, and therefore a living, from the sale of their products. Raising livestock is costly. Expenses include machinery, feed, fuel, veterinary care, drugs, and supplies. Livestock farming is also labor intensive. You

The traditional farm is owned and operated by families that make all decisions about the use of the land.

chatursunil / Shutterstock

must monitor the animals daily. There are no days off for weather, weekends, or holidays. You must constantly maintain facilities, equipment, and machinery. Sound land management principles, crop production, and storage, as well as a clear understanding of the nutritional needs of the animals are essential to successful livestock production.

What Is a Farm and Who Is a Farmer?

These may seem to be silly questions, as you probably have your own idea as to what constitutes a farm and a farmer.

The United States Department of Agriculture defines a farm as any place from which the owner produced and sold $1,000 or more of agricultural products. If you have a hive of bees in your backyard and you sell $1,000 worth of honey, technically you have a bee farm and you are a farmer.

Economic Impact of Farms

The 2007 Census data indicate that there are over two million farms in the United States, and they are responsible for producing $160 billion worth of livestock products each year (USDA-NASS, 2012).

Types of Farms

The majority of these farms (90%) are **family farms**. They are independently owned and operated. The farmer makes the business decisions about the land and how he or she will use its resources. He or she is responsible for his or her own purchases of animals, equipment, feed, and chemicals and must keep accurate accounting records of expenses and decide how to allocate assets.

The farmer decides *what* to raise and *how* to raise it. For example, let's look at beef cattle. He or she will determine the number of head that the land can support. The farmer usually grazes the animals on pasture and allows them to form their normal social groups. He or she rotates the cattle from pasture to pasture to avoid overgrazing. The farmer may breed the animals to produce calves or finish them (fatten them) to send to market. A farmer may used a portion of the land to plant crops to feed the cattle or may use that land as another source of income. This is a simple scenario, but you can see that an independent producer has control and decision-making ability over how he or she uses the farm. Because families often pass down family farms through generations, there are emotional ties to the land and the community, as well as a sense of stewardship. The goal is to nurture the land to provide a harvest, while protecting it for its continued use.

Corporate farming, also known as **factory farming** or **agribusiness**, is a different economic model than the family farm. The factory farm is one that a business with considerable financial assets owns. Corporate farms are often "vertically integrated," which means that the company owns everything that it needs to run the business.

To compare the two types of farming, let's say that Farmer Brown has fifty head of dairy cattle. She needs to buy feed for them, so she sets up a contract with the local feed co-op and buys their ration. This is good, because Farmer Brown is raising her cattle and supporting a local business, which is good for her community and its economy. Factory farms don't buy supplies locally. They own or make everything that they need to run their operation. The company controls every aspect of the business, including product distribution and marketing. This limits the surrounding community's economic growth, because the company does not purchase products and supplies from private businesses without prior arrangements.

Another crucial difference is the manner in which family farms and corporate farms raise the animals. Corporate farms generally keep animals in high numbers and in confined settings. Normal movement is very limited, and enclosures become contaminated with waste. Corporate farms bring the feed to the animals rather than animals having the ability to graze. The economic goal is to produce the most milk, meat, or eggs at the lowest cost.

There are controversies surrounding the corporate type of farming. Crowded conditions make the animals more susceptible to disease. The use of antibiotics and pesticides becomes necessary to control infections and parasites. The amount of feces and urine produced can become an environmental issue if it pollutes the land or local water sources, and animals can die in confinement areas without being noticed. Factory farming produces the majority of our food supply. The U.S. Department of Agriculture classifies 6 percent of all farms as factory farms, but these produce 75 percent of the available milk, meat, and eggs in the marketplace (USDA-NASS, 2012).

Family farm and factory farm sometimes merge into an agreement called a **contract farm**. These partnerships form with agribusiness companies to produce a specific product. The company determines what to grow and what price it will pay for the crop. The farmer still uses his or her land, but has entered into an agreement to provide a specific product at a predetermined price (USDA, 2012).

Advances in Livestock Technology

Technology has advanced agricultural practices for thousands of years. Humans have gone from hand-plowed fields to using oxen and horses to provide the labor. Then, by the mid-1950s, tractors outnumbered horses on U.S. farms. People have always looked for ways to decrease labor and increase efficiency. Technology has allowed us to increase productivity and quality in livestock by a number of methods. Following are some examples in the area of reproductive management.

craetive / Fotolia LLC

cheri131 / Fotolia LLC

Reproductive Management *Reproductive management* means the livestock producer will control the animal's reproductive cycle and plan its breeding. (This is a different approach from letting the animals decide when and whom to breed.) Both strategies offer advantages to the farmer, by producing animals with desirable traits and providing the ability to plan for their arrival.

    1. Timed breeding. It is an advantage to the cattle producer to be able to control when calves are born. The farmer can accomplish this by **synchronizing**

The transition from horse and plow to modern combine allows the producer to reduce labor and increase production.
Kadmy / Fotolia LLC

the cattle's reproductive cycles and breeding the entire herd within a limited time span. This method produces a calf crop at a predictable time, and calving season is shorter in duration.

2. **Selective breeding and artificial insemination.** Selective breeding is the planned breeding of animals for desirable traits. This reproductive strategy can increase the chances that animals will be born with the characteristics that the producer needs to improve the herd. For example, some consider horned cattle dangerous to humans, other animals, and sometimes to themselves. Horns are a formidable weapon. To decrease the possibility of injury, cattlemen dehorn bulls at a young age, usually when they also castrate them. One solution to this problem is to breed horned cattle to others that are "polled." Polled animals are born without horns. The polled gene is dominant and results in producing a naturally dehorned animal that is safer to house and work. Natural breeding or **artificial insemination** can accomplish selective breeding. Artificial insemination is a technique in which one collects semen from a mature male, then processes and stores it in liquid nitrogen. Breeders can use stored semen to inseminate the females when they come into heat. This is an advantage for **timed breeding** and also eliminates the need to keep a dangerous and unpredictable bull on the property. If Farmer Brown sees a bull that she thinks will help produce a bigger cow, she can artificially inseminate her entire dairy herd with semen from that animal.

3. **Embryo transfer.** Embryo transfer is a reproductive management technique whereby one transfers the fertilized eggs from a genetically desirable animal into recipient animals. This is an advantage because it produces more offspring of a specific genetic type than we would expect from natural breeding. The goal is to increase the animal's ability to produce viable eggs, which breeders fertilize and then implant. Typically, a cow has one calf per year. Embryo transfer technology allows multiple calf implants into recipients, thus increasing the superior animals' number of offspring.

Advances in Equipment and Machinery

A livestock operation's economic productivity increases with the number of animals that it raises. The farm's labor force limits this number. The replacement of manual labor with machines has greatly increased the herd size that an operator can manage.

If you have ever experienced milking a cow by hand, you have a good idea of how hard you must work for that milk. It is not economically possible to hand milk the number of cows on the average dairy farm. One person could not do it, and the costs for labor would be too high to hire sufficient workers to milk a

It is not economically possible to hand milk the number of cows on the average dairy farm.

Ammit Jack / Shutterstock

Automatic milking allows the dairy producer to reduce the time it takes for milking.
Diane Garcia / Shutterstock

herd. The automation of milking has improved the productivity and efficiency of this chore. The use of automatic milkers allows the cows to enter the milking parlor, have their udders cleaned, and have their milk harvested. Computer tracking records when the cow was milked and how much the animal produced. Automatic feeders dispense nutritionally balanced rations for the cows to consume during the process.

Up to now, our discussion of livestock has been limited to cows; however, technological advancements have increased our ability to raise many other species successfully and economically. The poultry industry produces over 8 billion broilers per year. Large farming operations raise many of these chickens in production houses where the density of animals can reach into the thousands. Technological advancements in ventilation, temperature, and humidity control conditions that otherwise might promote disease. Automatic feeders and a water system keep food and water sources clean and reduce manual labor. Waste management systems help process the enormous amount of fecal material produced by the birds.

Regulatory Agencies

You now see the significance that animals have on our economy, our culture, and our quality of life. To protect our food supply, and to ensure that we treat animals humanely, there are a number of federal agencies that oversee the use of animals in agriculture.

Animal and Plant Health Inspection Service (APHIS) The Animal and Plant Health Inspection Service (APHIS) is a branch of the U.S. Department of Agriculture, and its job is to promote and protect sound agricultural practices and safe products. One of its central goals is to protect livestock production from pest and diseases and to regulate the use of biotechnology that may threaten animal health. APHIS also supervises livestock imports and exports and monitors the health of animals entering at U.S. borders (APHIS, 2012).

Food Safety Inspection Service (FSIS) As you learned in Chapter 1, the Food Safety and Inspection Service is a division of the USDA and is responsible for the inspection and accurate packaging and labeling of meat, poultry, and egg products. Another important FSIS function is to ensure compliance with the **Humane Methods**

Large poultry operations produce tons of waste per year. Waste management is a key priority.
Tomas Sereda / Shutterstock

of Slaughter Act (HMSA). The HMSA requires that handlers of all animals entering slaughter facilities treat the animals in a humane manner. There are approximately eight hundred FSIS-inspected livestock slaughter facilities and three hundred poultry slaughter plants in the United States. The FSIS employs veterinary personnel to inspect theses facilities to monitor compliance (FSIS, 2012).

Environmental Protection Agency (EPA) The **Environmental Protection Agency** is involved in livestock production when animals are confined in large numbers. Confinement Animal Feeding Operations (CAFOs) are factory farms where animals are confined for more than forty-five days per year and have no access to grass or other forage in their confinement area during a normal growing season. These animals produce large amounts of urine and feces that can leach into the soil and contaminate groundwater. To regulate animal feces disposal, the EPA requires that every CAFO apply for a permit. The CAFOs must comply strictly with regulations that describe how they should handle manure and wastewater to avoid polluting water sources (EPA, 2012).

Food and Drug Administration (FDA) The use of antibiotics, hormones, and antiparasitic drugs is common in livestock. When producers use such substances, they must carefully record and monitor them as part of herd health management. The beef, dairy, or veal producer is responsible for holding animals from milk production or slaughter until the appropriate **withdrawal** time has passed. The Food and Drug Administration determines withdrawal times for those drugs that leave residue in milk or meat. The USDA-FSIS randomly tests animals at slaughter for drug residues. When an inspector finds meat or milk with drug residues, he or she traces them back to the producer. The USDA-FSIS may assess monetary penalties or revoke a farm's Grade A status (FDA, 2012)

Breed Associations

Breed associations are not federal agencies, but they do play an important role in improving livestock and promoting better health and husbandry procedures. Associations consist of members that have a special interest in a particular type of animal. The club's purpose varies according to the membership goals. Some focus on improving the animals' physical features for shows and competition or improving production while cutting costs. Associations sometimes keep extensive records that focus on the best methods and practices for raising the subject breed. For example, farmers raising Duroc pigs might want to join the state pork producers' association. Members will be interested in the same relevant topics: husbandry, reproduction, disease control, and economic concerns.

   The list below provides examples of breed associations:

   National Livestock Producers Association

   Poultry Science Association

   National Aquaculture Association

   American Dairy Goat Association

   American Sheep Industry Association

RESEARCH AND BIOTECHNOLOGY

Man has used animals in research and biotechnology for many centuries. Yes, centuries! The earliest Greek physicians practiced procedures and surgeries on animals before trying their techniques on humans. Since those early experiments, animal models have contributed to virtually all medical advancements.

Breed associations focus on producing desired traits for appearance, performance, and productivity.

Anke van Wyk / Fotolia LLC

What Is Research?

Research is the scientific method by which we answer questions and solve problems. In your own way, you have conducted research experiments that have an impact on your life. You try different laundry products to see which cleans best. You change ingredients in recipes to accommodate your taste. You drive different routes to school to see which one is shortest or which has the least traffic. A research project begins with a question, then proceeds to take the necessary steps to answer that question. Medical, veterinary, and biotechnology frequently use animals in the process of answering questions about how disease develops and how we might treat it.

A USDA report indicated that researchers used over one million animals in research in 2010. Researchers obtain animals from suppliers who meet housing and nutrition government standards and have provided appropriate veterinary care and husbandry (USDA, 2012).

What Is Biotechnology?

Biotechnology is the process of modifying living organisms. You may be surprised to know that despite its high-tech name, it is an old science. We have used the process of fermentation for thousands of years for brewing and baking. Although these are simple examples, they do illustrate how we have used biotechnology to our human benefit for a long time. Today, our understanding of DNA and gene sequencing gives us the ability to modify a plant or animal's genetic code. We can modify the genes that are responsible for milk production in cattle and pest resistance in crops or use bacteria to make human insulin. This type of science has enormous potential for improving agricultural production and drug therapy, but also poses many unanswered questions.

Biotechnology has contributed to animal health care as well. There have been many advances in veterinary medicine over the past several decades that allow us to better serve our patients. The development of antibiotics, less invasive surgical techniques, pain management, and diverse diagnostic tools are a few examples.

The production of pest-resistant plants is controversial. What are the long-term effects of genetically modifying our foods?

Zeljko Radojko / Shutterstock

Technological Advances

Transgenics Biotechnology has also opened the doors for improvements in animal and human health through genetics. One of the methods of manipulating DNA is **transgenics**. The process of transgenics involves the transfer of DNA from one organism to another. The benefit of this technique is to allow cells to perform processes they could not before. We call an organism that has had a newly inserted gene a **genetically modified organism (GMO)**. Currently, the vast majority of GMOs in use are plants such as corn. The modifications increase the plants' growth and natural resistance to pests.

The potential uses for this technology are staggering. Scientists are able to insert genes into cows to increase their lean muscle mass or to alter their milk to have the same properties as human breast milk. Scientists can genetically modify animals to produce human insulin, or to produce less phosphorus in their manure, which is better for the environment.

Gene Knockout Technology Just as we can introduce genes to DNA, we can also remove them. **Gene knockout technology** inactivates a gene in a DNA sequence. Once we remove a gene, we can tell what function it has. Because mammals share many of the same or very similar genes, gene knockout experiments can be used to develop an understanding of the human body. Exciting advances are possible, as gene knockout technology may also play a major role in providing a readily available source of organs for transplantation. For example, the pig heart is very similar to the human heart, but we cannot use it for transplantation because pigs make a protein that causes humans to reject their tissue. If we can remove the gene that makes the protein responsible for rejection, then we can transplant pig hearts into humans.

Somatic Cell Nuclear Transfer You might know the process of **somatic cell nuclear transfer** by the more common term **cloning**. The first successful mammalian clone was Dolly, the sheep that resulted when scientists harvested DNA from one sheep and implanted it in another. It is now possible to clone cattle, pigs, goats, horses, mules, cats, rats, and mice. This technology could be very useful for greatly increasing animal efficiency for agricultural purposes. Let's go back to livestock production and consider the average dairy farm. Most cows on the farm will produce an average of fifty to fifty-five pounds of milk a day. A few cows will make more, some less. If we could clone the top producing cows, then every animal on the farm would maximize milk production on the same amount of feed and space, resulting in a better profit.

Even though the results of cloning might be very beneficial, producers do not widely use the process yet. This is still a manual procedure that takes place under a microscope, so scientists fertilize each egg individually. There is often a high mortality rate among the cells. Until technology advances in the future so that the process is more economical, is faster, and has consistently repeatable results, somatic cell nuclear transfer will likely not be in common usage.

The Controversy

While animal biotechnology continues to be the cornerstone of research, it is also the focus of controversy. This debate focuses on two topics:

1. Is it ethical to use animals for research purposes? If it is, what species can we use and what types of research can we do? Is it OK to use rats and mice, but not dogs? Is it acceptable to conduct nutrition studies, but not surgical procedures? Opinions and emotions vary greatly on this topic, and both sides are well represented. There are powerful and vocal animal rights groups that would like to see the use of animals in research abolished. On the other side of the debate are federal agencies and privately owned companies that feel that animal models are the only acceptable way to conduct research.
2. Many people feel the use of genetically modified foods is "unnatural" and potentially unsafe. There may be consequences to modifying plant and animal food sources at the genetic level that we cannot yet determine. What would happen if herbicide-resistant corn transferred that characteristic to weeds? What might happen if the proteins that we consumed from meat have been altered? Would humans still be able to process them at the cellular level?

SUMMARY

A number of governing bodies and organizations whose roles are to protect animal and public health regulate the animal industry. Regulatory agencies exist at both the state and federal level, while the majority of animal advocacy groups are run privately.

The U.S. Department of Agriculture regulates animal welfare and food safety at the federal level. It is the USDA's responsibility to create and enforce government policy on farming, agriculture, and food. This encompasses animals that we use for food and labor. Since its creation in 1862, the USDA has passed many pieces of legislation regarding animal treatment and safety, including the Humane Slaughter Act of 1958, which insures that we slaughter as humanely and painlessly as possible the animals we use for food. The USDA also oversees all food processing plants, ensuring that we follow public health safety regulations.

Below the federal level, most laws related to animal welfare, health, and safety are left to the individual states to decide. Every state has a veterinary medical board that determines the "standard of care" that veterinary hospitals must provide. Local ordinances will regulate and enforce issues related to animal ownership, such as how many animals an owner may legally keep in a home and what basic provisions they must provide.

In the private sector, there are numerous groups that are devoted to protecting animal rights. Organizations such as the Humane Society of the United States, People for the Ethical Treatment of Animals, and the American Society for the Prevention of Cruelty to Animals dedicate their efforts to representing animals and monitoring how the animal industry and private ownership uses them (go to the HSUS, PETA, and ASPCA websites for more information). These organizations focus on raising public awareness in the areas that you have been reviewing—intensive farming for food and textiles, the use of animals in research and entertainment, and companion animal welfare.

Our way of life and the constant efforts we make to improve our standard of living is what drives and supports the animal industry. We want pets that are part of the family, food that is abundant and affordable, and medical advancements that help diagnose and cure diseases. Animals serve an enormous range of functions in our world, from recreation and entertainment, to food production, to the development of safe and effective drugs for humans. In order to meet these needs, the animal industry is constantly growing and evolving.

TEST YOUR KNOWLEDGE

1. In what ways do you personally support the animal industry? Make a list of three columns, labeling them *companion*, *livestock*, and *research*. In each column, place the name of a product or service that you personally use. Consider the list carefully. It is easy to overlook items that we use everyday, such as pharmaceuticals. (Note that researchers may have used animals as a component in drug testing before the FDA approved those particular medications for human consumption.)

2. Referring to the list that you made in question 1, for those items that have a fixed price, estimate what you spend on your companion animal for a year (example: 40 lb. bag of dog food a month at $25.00/bag = $300.00). Don't forget items such as toys, collars, leashes, veterinary care, licenses, grooming, and boarding. What do you spend every year on your pets?

3. Part of your job as a veterinary health professional is client education. How would you educate a client who is looking to buy a new puppy? What advice can you offer about purchasing a puppy from a reliable source? How would you advise your client on selecting an appropriate pet for his or her particular lifestyle?

4. What potential impact does exotic animal ownership have on public health and safety?

5. List three reasons why animal companionship is beneficial to human health.

6. Discuss the differences between family farm, corporate farm, and contract farm. Which farming operation provides the majority of food to the U.S. market?

7. What role does the Environmental Protection Agency (EPA) play in agribusiness?

8. What is selective breeding? Explain a benefit of selective breeding over natural breeding.

9. The process known as transgenics involves
   a. inactivating a gene.
   b. transferring DNA from one organism to another.
   c. the process known as cloning.
   d. a process for artificial insemination.

10. The government agency that tests meat for drug residue is the
   a. FDA.
   b. EPA.
   c. USDA.
   d. CAFO.

BIBLIOGRAPHY

APHIS. 2012. "Home Page." Accessed June 1, 2012 from the Animal and Plant Health Inspection Service.

APPA. 2013. "Pet Industry Market Size & Ownership Statistics." Accessed July 28, 2013 fro the American Pet Products Association website.

APPA. 2013a. "National Pet Owner's Survey." Accessed July 28, 2013 from the American Pet Products Association website.

APPA. 2013b. "National Pet Owner's Survey." Accessed July 28, 2013 from the American Pet Products Association website.

CDC. 2012a. "Healthy Pets Healthy People." Accessed July 25, 2012 from the Centers for Disease Control and Prevention website.

CDC. 2012b. "Diseases from Reptiles." Accessed July 25, 2012 from the Centers for Disease Control and Prevention website.

CDC. 2012c. "Monkey pox." Accessed July 25, 2012 from the Centers for Disease Control and Prevention website.

EPA. 2013. "Animal Feeding Operations." Accessed May 31, 2012 from the Environmental Protection Agency website

FDA. 2013. "Animal and Veterinary Home Page." Accessed June 1, 2012 from the Food and Drug Administration website.

FSIS. 2013."Humane Slaughter." Accessed May 30, 2012 from the Food Safety and Inspection Service website.

USDA-NASS. 2012a. "2007 Census of Agriculture." Accessed May 13, 2012 from the United States Department of Agriculture-National Agriculture Statistics Service website.

USDA-NASS. 2012b. "2007 Census of Agriculture." Accessed May 13, 2012 from the United States Department of Agriculture-National Agriculture Statistics Service website.

USDA. 2012. Accessed May 13, 2012 from the United States Department of Agriculture Economic Research Service website

USDA. 2012. "Annual Report Animal Usage by Fiscal Year." Accessed November 26, 2011from the United States Department of Agriculture website.

Learning Objectives

At the end of this chapter, you should be able to:

- Improve workplace relationships by developing good communication skills.
- Refine your professional conduct with coworkers and clients.
- Use role playing to practice situations that represent common clinic interactions.
- Develop confidence in your interactions with clients and coworkers.
- Manage difficult clients in a professional manner.
- Employ problem-solving strategies to diffuse and diminish confrontations.
- Utilize professional behavior to meet and exceed the expectations of your employer and coworkers.

5

Developing People Skills and Work Ethics

The author gratefully acknowledges the Chapter 5 contributions of Ray Higley, MAEd.

Chapter Outline

INTRODUCTION

Soft Skills

How much of your time will be spent working? If you start at the age of eighteen and work a typical forty-hour week five days a week until you are sixty-five years old, you will have spent 97,760 hours of your life in employment. For forty-seven years, you will spend one-fourth of your time in the workplace. It is time to make an investment in the "soft skills" that lead to success in any organization. Soft skills are different from technical skills. Technical skills embrace specific tasks such as placing an IV catheter, preparing for surgery, and performing dental prophylaxis. With time and practice, technical skills improve. Soft skills are what we commonly call "people skills," and for many students they are the most difficult to learn. Interactions with your animal patients will not require the social civility that your clients will expect. You must learn methods to professionally communicate and preserve good relationships with coworkers and extend the best impression to your client. The task of caring for animals is a complex job that requires you to master many competencies. This chapter will help you shape your behavior so that you learn to develop exceptional people skills.

People who are drawn to the veterinary profession often state that they prefer animals to people. Does that describe your feelings? Can you take one look at a dog's tail and know what behavior to expect? Can you determine a horse's thoughts by the position of the ears? Those signals are easy to interpret for someone who is attuned to animals. Now ask yourself, can you diffuse an angry client who is dissatisfied with the outcome of treatment? Can you resolve a conflict with a coworker without it escalating to a verbal or physical fight? This is a good time to think about the realities of working in a veterinary setting. If you are considering working with animals so that you can avoid working with people, this is a misguided concept that you must quickly put to rest.

As a veterinary team member, you have a double challenge. Not only must you provide excellent, competent care for your patients, you must earn and maintain the trust of your clients. You have to conduct open, honest, professional relationships with your coworkers and colleagues. Earning this trust begins with establishing a solid foundation of personal characteristics that say to people, "I am honest and hardworking. I will do what I say I'm going to do to the best of my ability. When I need help, I'll ask for it. When someone needs help from me, I'll give it." You have to convey this message with your actions and your words.

If you find the animals easy and the people difficult, this chapter will make your interactions with humans easier and more enjoyable.

As you can see from this and the following 5 photos, animals are expressive.
gillmar/ Shutterstock

Goldution / Shutterstock

Marcel Jancovic / Shutterstock

Makarova Viktoria / Shutterstock

Sinelyov / Shutterstock

You can clearly see from the difference in expressions and body language the messages these animals are conveying. The small terrier is running with his tail up, ears pricked, and an alert expression. The Chihuahua has a cowered body posture and tucked tail. The first horse has a relaxed eye and his ears are pricked forward. The second horse is charging with his ears pinned back. The first kitten has a curious expression and his head slightly cocked to diffuse tension. The second kitten is clearly signaling a warning with crouched posture, flattened ears, and teeth displayed.

Vishnevskiy Vasily / Shutterstock

You will need to earn the trust of your client with your actions and gestures before she will be willing to let you work with her pet.

absolut/ Shutterstock

COMMUNICATION

Communication in all its forms is about the relay of information. Communication can be conscious or subconscious, verbal or nonverbal. We communicate information in a variety of ways. When we speak, we can enhance what we say by tone, inflection, and volume. Communication can also be written or inferred from body language. You are receiving and sending these messages every day as you assemble and interpret body-language messages to gather the information you need to interact in an appropriate way.

Communication is essential in all relationships so you have a clear understanding of what people mean and what they expect of you. Everyone knows the frustration of not fully understanding what someone is saying and in turn not having the ability to respond or meet a need. Relationships break down from poor communication. In a professional setting it becomes critically important that the lines of communication between coworkers and clients be clear. In veterinary medicine, 40 percent to 50 percent of all malpractice suits filed are the result of poor communication (Bayer Animal Health, 2012).

It is crucial for the veterinary team to communicate. The Bayer Animal Health Communication Project study results indicated that the number one reason why people choose and stay with their veterinary hospital is the way that the practice treats them and their animals. Clients seek a relationship built on trust. They trust that you will listen, that you will provide excellent care for their animal, and that you will communicate in a way that is easy for them to understand.

It takes more than technical expertise, great skills, or a clinic full of the latest diagnostic machines to accomplish this. Interpersonal communication skills are every bit as important as any piece of equipment. You will have to engage in many types of discussions with your clients and coworkers, and some will be very difficult. You will have to address topics with your clients that will make you uneasy if you do not have the proper communication tools. Issues such as money, end of life, and euthanasia are common in veterinary practice, and even under the best of circumstances these issues are challenging.

First, look at the list of personality traits that are essential for building positive relationships both in and out of the workplace. Then, let's take a look at how to improve our communications by comparing the ways in which we relay information to one another.

Nonverbal Communication

Body Language Imagine that you are escorted into an empty room. You are told to take a seat. Your only instructions are that you are not allowed to speak to anyone who may join you. You take a seat and wait a few minutes. The door opens and three more people walk in and sit down. What do you do in these first few minutes while the group awaits instructions? You size up each other.

It is no secret that most communication between people is nonverbal. This means that you are constantly giving physical signals, conscious or unconscious, that people observe and interpret to assist them in judging the meaning of your behavior and the elements of your personality. This happens so fast! It only takes a second or two for you to have an idea of what someone is like before you ever exchange a single word.

We often refer to nonverbal communication as "body language." Reading body language goes hand in hand with listening and adds significantly to human understanding. A client's facial expressions, hand gestures, posture, and stance are vital signals that can reveal important details prior to your formal introduction. Body language often gives away our intentions and feelings, because it is usually not consciously controlled. You are most likely able to interpret body language intuitively.

Mrs. Jones has come into the clinic. She is sending you a message that she is worried about her dog before she gives you any information—you can tell by her closed posture and her furrowed brow. She is sitting with one arm crossed over her body, the other over her mouth; her head is bent, and she is looking up at you for reassurance.

Mr. Smith in exam room 2 is not happy that this is his third visit to the clinic with Jake, his Labrador retriever. When you walk in, he has his receipts in his hands,

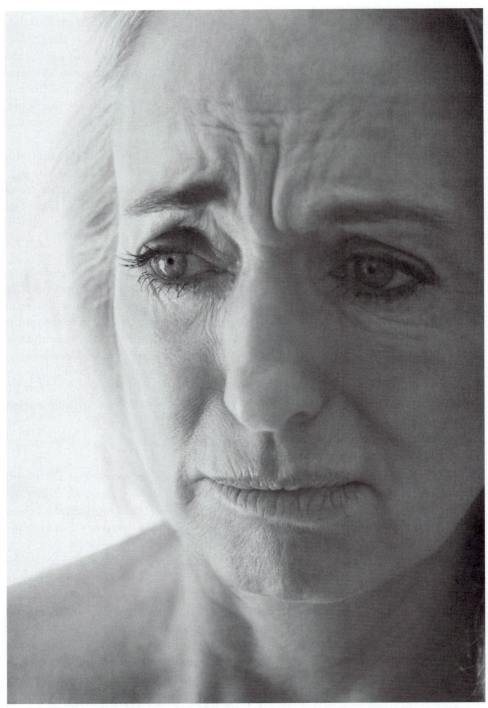

Worry is communicated on Mrs. Jones's face. Notice her strained expression and furrowed brow.
Monkey Business Images / Shutterstock

he is standing with his feet are apart, and he greets you with a hard stare. You know immediately that Mr. Jones is angry.

Remember, you are sending signals of your own. Never neglect the message that *your* body language is sending to the client. It expresses information about your mood, health, and even your level of training and competency. Do not diminish the importance of this information. Your clients are just as able to read your nonverbal

While Mr. Jones's brow is furrowed, his expression communicates anger because it is accompanied by a hard stare.

Felix Miozznikov / Shutterstock

cues as you are theirs. They will instantly assess your attitude. Are you genuinely glad to see your patient, or is this just another vaccination appointment? Is your body language telling your client that you are tired and would rather not be bothered? Do they sense arrogance or superiority in your attitude?

Your challenge is making sure that you modify your own gestures to convey positive messages to your clients and coworkers. All of us enjoy a relaxed set of standards with regard to our body language when we are with family and friends. Part of your training will involve modifying, and in some cases eliminating, postures and gestures that are not appropriate for the workplace.

Table 5.1 lists common elements of body language that require some practice to become second nature. Enlist a classmate or friend to help you. Take pictures or

TABLE 5.1
NONVERBAL COMMUNICATION

Characteristic	Desired Behavior	Avoid or Eliminate
Posture	Standing straight and relaxed	Slouching, arms crossed over chest, lowered head
Eye contact	Direct eye contact	Hard stares, unfocused eyes, looking to the sides or at the floor

StockLite / Shutterstock

Jorge Casais / Shutterstock

Arms and hands	Open or at your side	Clenched fists, fidgeting, hands on hips, crossed arms

karamysh / Shutterstock

Smile	Genuine and open, including eyes	Pasted, artificial

Danny Hooks / Fotolia LLC

even make a video recording of your normal appearance, and use it as a guideline to help you develop a professional demeanor.

It is important that you do not fall into the trap of judging someone solely based on his or her body language. Remember, people most effectively understand body language when they observe and interpret it in context. It is normal for otherwise-confident people to be scared, anxious, or even aggressive in veterinary hospitals if they think that their pet's health is compromised. Many people are nervous in anticipation of needles, blood, or medical procedures, and clients can become unnerved if they witness their pet in discomfort or receive an unexpected diagnosis.

To better understand this, try to think of a situation in which you are completely out of your comfort zone. It might be at your physician's office, at an auto repair business, or when trying to deal with advanced technology. When you are in unfamiliar situations, you take cues from people who are there to help you. Try to remember the last time someone made you feel welcomed and relaxed. Duplicate that behavior for your clients.

Eye Contact Making and maintaining eye contact with fellow humans has an enormous impact on how people interpret your behavior. It is different than the methods you will employ to gain trust in your animal patients. The types of shared eye contact have great variability between cultures and local customs, and you must recognize this. In general, vacant looks, hard stares, or an inability to look someone in the eye while speaking makes for a poor impression. And we usually interpret eye rolling as a sign of disrespect. Comfortable eye contact is relaxed and is often accompanied by a smile, nodding, or raising the eyebrows in acknowledgement.

Personal Space Animals and people have a personal space. This is a boundary that they maintain as a psychological comfort zone. Entering a personal space is usually a function of relationship. Family and friends are allowed closer access, while mere

This teenager is displaying disrespect by rolling her eyes.
Elenathewise / Shutterstock

acquaintances and the general public are more restricted. Be aware that ethnic and cultural differences will alter these limits. Someone might interpret an encroachment on his or her personal space as threatening. Be mindful of maintaining appropriate distances so that people do not misinterpret your intentions. If it does become necessary to breech a comfort zone, ask permission or communicate it to the client or coworker before any action.

Mirroring Mirroring behavior is a tool that you can learn and use to establish trust and understanding. Mirroring is the process of observing someone's positive body language as well as their postures and gestures or facial expressions and then duplicating that behavior. You do this unconsciously all the time, but it can be an effective way of making people relax and feel connected. They feel that they like you because you are reflecting back their own behavior. The first step in learning to mirror behavior is to observe what your client or coworker is doing. Are they seated or standing? Are they leaning forward or backward? Do they have their hands in their lap or in their pockets? Are they speaking in a loud tone of voice or a soft whisper? Mirroring says, "Relax . . . I am like you."

You can practice mirroring with someone who you do not know well, a new friend, or an acquaintance. See if the flow of the conversation and interaction seem easier and more pleasant. Do not mirror negative body language—this only escalates tension and can lead to arguments and bad feelings. Remember, we use mirroring to enhance pleasant interactions and make people feel at ease. It should never appear as "monkey see, monkey do."

Verbal Communication

Deducing the meaning of common body postures and facial expressions and mirroring positive body language is a good starting place for understanding nonverbal communication. Verbal communication is the second component. This is where you might have to modify your habits for proper workplace behavior. Verbal communication at home and with friends is vastly different from the phrasing and speech that we use with clients. You must take care to eliminate annoying and counterproductive habits such as interrupting, dominating the conversation, or adding personal information and experiences. You may also need to learn to modulate the tone and volume of your voice. Let's examine some useful techniques to practice good verbal communication.

The Art of Listening Before we begin speaking we must learn to *listen*. Our profession is challenging because our patients lack the ability to describe what they are experiencing. In part, we rely on their owners to communicate any concerns and observations. The pet owner shares our common desire to provide these special companions with a quality life. In order to gather the best information on the patient's condition, we must learn to listen to those who know them the best— the owners. An owner's level of understanding will vary dramatically. You must develop a sense for listening carefully to what the owner is saying, and then apply your training.

Make a habit of listening to every detail that the pet owner provides, and develop a strategy for interpreting this information. Writing down information as the owner describes it is a common way to keep facts straight and organize your thoughts. Make sure that while you are listening, you devote your full attention to the person who is speaking. Do not let your attention wander or be distracted by other events taking place in the hospital. Maintain eye contact, and give verbal cues signaling that you are engaged and concentrating. Keep your comments to yourself

This doctor is actively listening to her client. She is making eye contact and is leaning forward.
Andresr / Shutterstock

until it is your turn to contribute or ask a question. Never interrupt someone in the middle of a sentence or finish the person's thought. It gives the impression that you are impatient. With experience, you will hear many of the same patient histories day in and day out. Listen to each client as if he or she is the first one to have ever shared this information. Make him or her feel as though he or she is the most important person and patient in the hospital. As your ability to listen improves, you will also be able to process the body language and eye contact signals simultaneously. Together, they will combine to give you a reliable assessment of behavior. If you practice your skills in a variety of settings, the habit will become second nature and serve you well in your professional service. It is a great skill. Who doesn't love having someone simply *listen* to them?!

Another useful habit that contributes to excellent listening skills is *using the person's name* when addressing your client. When discussing the patient, refer to the animal by the name and correct masculine or feminine pronoun. Owners mentally and emotionally relate to the gender characteristics of their pets, and it is an important component to your good client service skills to pay attention to these details. This will instill confidence in your client and may place the animal at ease. "Mr. Hatchett, I noticed that Gash was limping on his left hind leg as he was walking across the parking lot. How long has this been going on?" Addressing a client and patient by name will gain favor in most situations.

Many clients are anxious about relating to you what seems to be ailing their pet, particularly if it is of a sensitive nature. Allow them to tell their story. Most people are a bit reluctant to refer to body functions or certain anatomical areas by proper names. They may use terms that diffuse their embarrassment from lack of understanding. Listen carefully to what they say and how they describe or demonstrate the animal's behavior. Some descriptions have regional or cultural components, so when you listen to your clients, it is important to understand the precise meaning of their words as it relates to the animal's health; this will allow you to record an accurate history.

Paraphrasing When the pet owner has finished, start investigating the exact meaning of his or her words. An effective lead-in for questioning is to **paraphrase** what her or she says. "If I understood you correctly, Mr. Hatchett, you were splitting fire wood when Gash dashed between the axe and the log, and the axe unfortunately struck Gash's left hind leg. Is that correct?" This line of questioning will solidify the eventual diagnosis and treatment. It will also better prepare you to relay precise information to the veterinarian. In addition, your client will have gained confidence in your professionalism because you were willing to listen rather than make any premature, misinformed assumptions.

Wait for Verification Finally, seek confirmation from your client that what you heard is consistent with what he or she described. The art of listening pivots around the ability to translate what you have heard into meaningful language for the veterinary caregivers with whom you work in the clinic. Therefore, ask your client for confirmation, and wait for him or her to acknowledge that your version is accurate.

This will allow the client to further elaborate on the symptoms or alter information that you may have misinterpreted. Most of your clients are not trained to the extent that you and your coworkers are, and developing a clear line of communication will not only enhance your relationship with the pet's owner, but also provide the best information for starting treatment. Take this information with you when you begin your formal examination of the pet in question.

GRAMMAR, SPEECH, AND SPELLING

Many students struggle with proper grammar, speech, and spelling, but it is an important component of your professionalism. You will be expected to phrase your sentences properly, pronounce terms correctly, and spell precisely. You can do it,

Mr. Hatchett, just before the accident.
Sergey Kamshylin / Shutterstock

but you may have to work at devoting some time to practice and ridding yourself of some bad habits. In the workplace, it is best to avoid slang and vulgarity. There are no circumstances in which profanity is acceptable. Your grammar is important too and says a great deal about your education. Using poorly phrased sentences creates a negative image for you and the workplace. Some of these grammatical errors are so common that you may not even realize that they are incorrect. Phrases such as

"I didn't see nothing" or "we was on break" can be as offensive to the ear as a trash heap is to the eye.

Incorrect	Correct
We was . . .	We were . . .
I seen/we seen	I saw/we saw
brang	brought
boughten	bought
The floor needs mopped.	The floor needs to be mopped.

Proper spelling is difficult for many people. In medical disciplines it is critical. You will be communicating with your fellow professionals, making notations in medical records, and creating informational material for clients. Many drug names are very similar, and you must spell them properly to avoid a clinical error. For example: Hydroxyzine is a commonly used antihistamine; hydralazine is a drug used to treat high blood pressure. Insulin is a hormone that lowers blood glucose; inulin is a plant starch. Enalapril is a drug used to treat high blood pressure; Anipryl® is used to treat canine cognitive dysfunction.* You can see from these examples that correct spelling communicates the precise treatment plan for the patient.

YOUR PROFESSIONAL APPEARANCE

Dress the Part

You are in training for a professional career as a veterinary technician. Nearly every profession has established expectations when it comes to appearance. Every aspect of the dress code is developed for a reason—from the design of your scrubs to the avoidance of dangerous jewelry. Those who have been practicing animal caregiving activities over many years have passed along several standards in appearance that you should not take lightly. Some are for your safety, and some are to convey confidence in your clientele, but they are not for frivolous reasons. How you present yourself to the public will often determine your eventual success or failure in the veterinary field.

The dress codes for practice vary but have safety as the primary concern. If you have the option to select your own work clothing, do so with much less thought toward fashion, and more on movement. You will have a physically demanding job. Your clothes or scrubs will have to allow you to move freely with a struggling animal in your arms. At the same time, they should not be so loose fitting that they drag on the floor or expose excessive skin. Pants that are too long are a hazard. You do not want to grind dirt or contaminants into the bottoms of your pant legs. Over time they will become soiled, torn, and stained, and will therefore require replacement. Keep in mind that your work clothes are also your uniform. They are a significant part of the first impression that your clients will remember when evaluating your overall professional worthiness, so dress the part.

From the way you wear your hair to how you display body art or use jewelry, clients are judging your appearance. Your personal appearance may not directly affect your technique, but it will send a message to your client. Keep your jewelry choices subtle and with safety in mind. A torn ear lobe is a high price to pay for wearing large

---

*Canine cognitive dysfunction: As dogs age, they may develop a condition in which dopamine, an important neurotransmitter, decreases. This decrease leads to a loss of cognitive function. Dogs may develop separation anxiety, have house soiling issues, or appear lost and confused in familiar environments.

A sloppy appearance at work conveys an attitude of carelessness.
Amy Wolff

or dangling earrings. Safety is first and foremost. Many clinics adhere to dress codes that will describe what is and what is not acceptable in the workplace. There may be limitations on hair color, visible body art, or piercings. The decisions regarding the acceptability of these types of adornments will rest with the veterinarian or office manager and be influenced by the region where you work and local custom.

There is precious little time in the day of a veterinary technician when you will be off your feet. You will learn quickly how a sore back or feet will hinder your ability to work effectively, and chronic pain influences your willingness to extend yourself to others. Select comfortable footwear with adequate support that will not hinder you in your daily tasks. For some who work with large animals, this may require the need for a boot with reinforced toes. When you work in a clinic, always be aware of how your footwear appears. Keep shoes clean and presentable; those with stains or holes are unacceptable. Many employees have a pair of shoes that they only wear in the clinic. This helps avoid excessive wear and tear and also prevents disease transmission. You do not want to bring home to your own pet a bacteria or virus on your shoes. Footwear is not an item of dress where it pays to spare expense; the extra money you spend on good shoes or boots is an investment that will be returned in the long term.

As a final thought on dress, do not assume that clients will overlook neglected hygiene. Perhaps your animal patients will forgive you, but they are not your paying customers. Their owners are. There is simply no reason to begin your workday appearing unkempt or having an unpleasant body odor. Restraining veterinary patients frequently requires two or more people to be in very close proximity to one another. There are plenty of disagreeable odors in veterinary practice. You should not be one of them. If you smoke, you need to be aware that nonsmokers are very sensitive to the odors that you carry on your clothing and breath. If you are allowed to take smoke breaks at work, freshen your clothing and use breath spray, gum, or a mint. Keeping small toiletry items handy—such as toothbrush and paste, mouth rinse, deodorant, hair brush, and a lint/hair roller—helps you stay clean and presentable.

There may be restrictions concerning visible body art in the workplace.
iofoto / Shutterstock

YOUR PROFESSIONAL BEHAVIOR

Positive Personality Traits

When you hear the term *personality trait*, it is a reference to the habits and patterns of your character and the foundation for your relationships. Positive personality traits are those that inspire trust, love, and respect. Negative traits are detrimental to relationships and lead to distrust or social impairment. There are many positive personality traits that you will work to build and improve over a lifetime. Within the framework of the veterinary setting, there are traits that are worth a special mention, as they are the cornerstone of the relationships that you build with your clients.

Honesty Your behavior and actions are based on truth and integrity.

Kindness Acting in a helpful manner without thought of personal gain (see *caring*).

Compassion The desire to stop or reduce another's distress (animal or human).

Dignity Having your needs recognized and honored; holding another person in esteem.

Empathy To share and understand another's feelings.

Respect Honoring and meeting a person's needs.

Caring To show interest and concern for the welfare of another without the intention of personal gain (see *kindness*).

Trust You are reliable and accountable for meeting your obligations.

Tact Methods that you use to avoid offending someone or hurting his or her feelings.

Every day you will be called upon to share these behaviors with your clients and patients. You must learn to communicate them in an appropriate manner. Polishing your professional behavior will help communicate your positive attributes to your clients.

Your Attitude

Beyond the decisions that you make for clothing, footwear, and accessories is the overall appearance that you present with your attitude. Your goal is to let all clients know that you appreciate them, hold them in high regard, and have as your number-one priority the care of their pets. All client interactions should begin with a sincere smile accompanied by eye contact. You should smile during phone contact as well, because that positive gesture is communicated in your voice. Smiling sends a universal message that every person who walks into your clinic understands. If the first signal that you send a new client is your smile, then you are on the right track. Remember, clients are customers at all times, but you can only refer to them as clients if they continue to return to you for future advice and care. Turning a customer into a client requires that you balance your knowledge of animal science and veterinary skills with an understanding of the relationship that develops over time with people. Your daily attitude must be consistently pleasant, regardless of developments that may be occurring in your personal life. A professional learns to maintain his or her level of caregiving, regardless of what may be taking place away from the clinic.

Let Your Passion Show Consider for a moment the time and effort you are investing in an education—not just in terms of contact with your instructors, classroom hours, or the cost of tuition. There are other costs such as hours of study time, both with fellow students and on your own. In addition, there is driving time to and from classes, not to mention the gas expenses. Costs of books, supplies, scrubs, and certainly the time that you are away from family and friends are part of the commitment that you are making. This level of sacrifice must require an incredible amount of passion. Without that drive, your goals would not have a chance to materialize.

There is a desire inside that is motivating you. At this point, it may seem like a long journey ahead to reach your goal, but you know it is there—otherwise, you

u20 / Shutterstock

The message of a smile is universal and occurs across species.
Eric Isselee / Shutterstock

Rosaline Chan / Shutterstock

would not be here. Keep focused, and let your imagination create for you an image of yourself working among peers who are qualified to deliver a full day of work. Let your passion show. You have earned it. Bring it with you and show it to every client and patient. Extend those feelings into your community as well. There will be ample opportunity for you to volunteer in various events that raise awareness of pet care and responsible animal ownership. Use your passion to advocate for positive changes. Volunteer at a shelter, join a professional association, supervise a wellness event, or join your state's emergency response team. There is a wealth of possibilities to help you turn your job into a *career*.

Confidence As a student faced with learning so many new tasks, it is common to feel overwhelmed and confused. Classes and training progress at a rapid pace, and it is often a struggle to keep up with all of the new information and skills that you are required to master. You are also beginning to accumulate significant knowledge about multiple animal species. Are your friends and acquaintances asking you questions about animal health when they find out that you are a student of veterinary technology? How confident do you feel in answering their questions or providing them with accurate information? Your confidence level will rise as you spend time practicing your skills. Every day you will learn new information and different ways to solve a problem. Your employer may ask you to assume responsibilities in the hospital before you feel you are ready. How can you build confidence so that you can best serve your clients?

One way to approach this problem is with **goal setting**. Set yourself one small reasonable goal and a time frame in which you will accomplish it. Do not limit yourself to physical skills only. You will devote much of your time to client education. You might set one of your goals as becoming the flea-control expert in your clinic. Soon you will find that when Mr. Hatchett returns with Gash to have his sutures removed, you can easily and confidently speak to him about the flea dirt that you noticed in Gash's hair coat and be able to make appropriate recommendations. You will find that you will need to devote a portion of your personal time to this type of learning. You will need to access all types of resources—books, magazines, online forums, and continuing education events—and, of course, the help of your coworkers. Remember, you are the expert! Your clients will expect you to know what you are talking about. People have easy access to information, and when they do their own searches, they will be pleased to find that you have given them accurate, up-to-date information.

Once school is over, you may be tempted to shelve the books and other resources that you have accumulated. It is important to remember that the real learning starts after school is over. Education is a foundation for lifelong learning. You are being given the tools to assist you in your commitment to care for animals. Part of your professional work ethic is to ensure that you always have current information regarding animal health care.

Check Your Ego and Your Temper With a year or two of experience under your belt, you will feel more capable and confident. The hesitance that you felt when you were first out of school has given way to the confidence of a competent professional. As you cross the line from the cautious, apprehensive student to the employee who now has many important responsibilities, you must modify another important aspect of your nature—your ego. Many people will envy your job and the everyday contact you have with animals. This is a good thing! Use their interest to educate and inform, but never pass yourself off as a know-it-all. Living things have a great capacity to heal themselves. As veterinary professionals, we just give them the chance to do that. The practice of medicine is truly a humbling experience.

Humility is maintaining a modest opinion of oneself. Never use your knowledge and education to bully or intimidate someone. Making someone feel guilty because he or she has not met a standard of pet care is not a good way to use your influence. No one wants to associate with someone who makes us feel bad about decisions or actions.

For example, let's revisit Mr. Hatchett and Gash. They are back for Gash's yearly vaccinations and physical exam. Mr. Hatchett requests a rabies vaccine only, because that is the shot that he needs for obtaining Gash's dog license. Mr. Hatchett has not followed your recommendations about using flea prevention, and Gash has a terrible case of flea bite allergy. A blood test indicates that Gash is positive for heartworms. When you see the results, you are furious! This poor dog has already suffered that terrible laceration and has had fleas for six months. You just want to scream!

Frustration may prompt you to approach Mr. Hatchett and tell him that if he had listened to you about giving the dog his heartworm preventative, this never would have happened to Gash. You think that Mr. Hatchett must be the *only* owner

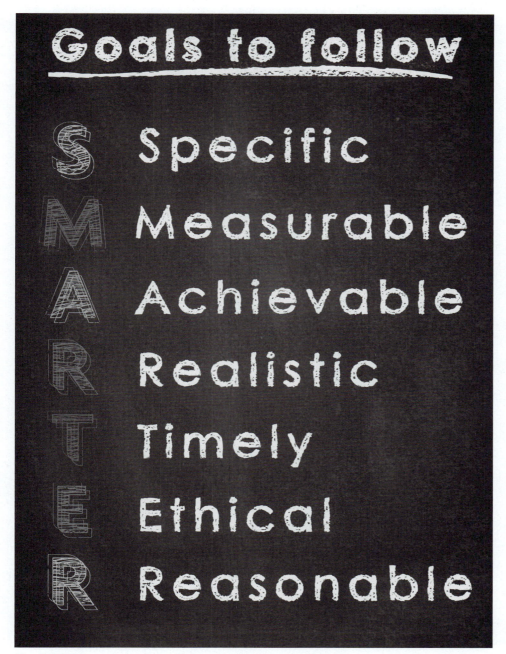

The SMARTER principle is a tool for goal setting. Use the guidelines to set specific goals that are within your abilities to achieve. For example, the goal of "being the best vet tech in my city" is unrealistic and can't be measured. How do you know if you have achieved this goal? A better approach would be to say, "I am going to earn a specialty certificate in dentistry from the North American Veterinary Technician Association (NAVTA)." This is a measurable goal.

Vaju Ariel / Shutterstock

on the planet who has not heard of heartworm disease. Before you lose your cool, stop and think. What will you accomplish by treating Mr. Hatchett in this manner? Certainly, if your goal is getting him to admit Gash to your clinic for treatment, this is not the best approach. Although the Hatchett family ignored your advice and neglected Gash, the proper way to present the information is in a calm and confident manner, with Gash's best interests in mind. Your must be patient and offer compassion. Compare these two approaches:

Johan Larson / Shutterstock

Negative: "Mr. Hatchett, we told you months ago that Gash needed flea and heartworm preventative. Now this poor dog has heartworms. If you'd have bought preventative when we told you to, this wouldn't have happened."

The only person who will feel good about this is you, and only temporarily. In the long run, you will regret your treatment of Mr. Hatchett, and he and his family will be reluctant to seek veterinary care if that is the experience they have in your clinic. It is doubtful that your veterinarian will advocate this approach. It is a lose-lose situation.

Positive: "Mr. Hatchett, I'm so sorry to hear that Gash's tests indicate heartworm disease. Dr. Newton has asked me to prepare an estimate for his hospitalization and treatment. I'll have it for you in a moment."

Take the time to answer any questions Mr. Hatchett may have about the diagnosis and the suggested treatment to remedy the situation. Support your veterinarian by preparing information, estimates, or consent forms. Treat Mr. Hatchett as if he were your best client. Who knows? Maybe he will be.

Maintaining Your Work Space

Now that you have begun to polish your professional appearance and attitude, let's apply those same principles to your work space. There are many areas in the veterinary hospital that all staff members will use. If the hospital is clean and organized, the people working there are less likely to be frustrated and overwhelmed with inconsistency and sloppiness. You can also experience a feeling of self-satisfaction from extending the life of your equipment, work space, and instruments through regular upkeep. There will be numerous daily chores in a veterinary hospital, many unpleasant. It is easy to ignore dishes in the break room sink, a dirty toilet, or a pile of stinky laundry. No one wants to clean up a mess that he or she did not create. It is best to put away petty feelings of "it's not my mess." The reality of practice is that everyone in the hospital is responsible for keeping things cleaned, organized, stocked, and in working order. If one of the staff members continually picks up after the others, that is an issue best resolved in a staff meeting or a discussion with a supervisor.

In general, other coworkers will clean up after you as much as you clean up after them. No one wants to come into a messy work environment. It will decrease your ability to care for your patients in an efficient and timely manner. If you constantly have to look for things, do not have the necessary supplies ready, or have to clear away items that have not been properly stored, in order to have a work space, soon you will be angry and frustrated.

A messy desk results in poor patient care. Records, reports, messages, and other documents are often lost or misplaced in a disorganized work space.

trekandshoot / Shutterstock

TAKE CARE OF YOURSELF SO YOU CAN CARE FOR OTHERS

Your Emotional Bank

Just like a savings account, each of us has an **emotional bank**. An emotional bank is the energy that you draw on to be able to meet the needs of others, be kind and caring, and be respectful of their feelings. As a veterinary professional, your emotional bank will need to be very full, because you will be dipping into your "savings" every day. It is important that you meet you own emotional needs so that you do not bankrupt your account. Your emotional bank deposits are rooted in the relationships that you have with family friends and coworkers, and are centered on trust. You can also make deposits by doing the things that you enjoy. When you make a deposit into your emotional bank, your sense of self-esteem and confidence grows. If your supervisor comments on the great organizing job you did in the pharmacy, or how well you documented your communication with Mr. Hatchett when he called to check on Gash, you have a deposit in your emotional bank. When a friend thanks you for helping him or her study, that becomes a deposit in the bank. On the contrary, if a client is rude or abusive, this is a withdrawal from your emotional bank. The goal is always to have more deposits than withdrawals in the bank. Excessive withdrawals from the bank lead to job dissatisfaction and burnout.

Look forward to a little personal time that restores your emotional well-being. Do the things that make you relaxed and refreshed so that you are able to give that positive energy back to your workplace. You might consider spending some time engaged in a hobby or pastime that has *nothing* to do with animals. Your goal is a healthy outlet for the stress, frustrations, and even successes at the workplace. Make those goals realistic and achievable. Planning for activities that exceed your time and financial resources creates more obstacles. Put your energy into productive activities. Exercise, reading, music, movies, and social and sports events are examples of activities that you might use to restore your emotional bank balance and sense of well-being.

Adopt a Mentor

No matter what aspect of veterinary medicine your career path follows, you will meet amazing and talented people. They will perform their jobs so skillfully that it will appear effortless. You will come to admire many of these people even though they might make you feel a little intimidated. Younger, inexperienced employees often need role models to help them understand professional behavior in the workplace. You will feel more comfortable, less stressed, and more willing to learn when you have a coworker's support. This is your opportunity to "adopt" a mentor.

A mentor is a person who takes an interest in your personal and professional growth. Mentoring is a generous gift. There is no monetary compensation or expectation of a returned favor. Mentors enjoy sharing their knowledge with people whom they recognize as having a genuine interest in learning and a desire to continually grow in the field. Mentors often recognize talents in other people that they do not see in themselves. You may feel incompetent and be inexperienced, but you are smart. You just need time and assistance. Mentoring is a great way to help a new employee with the transition from school to the workplace.

There is a great deal of difference between theory and practice. A mentor can demonstrate how things are accomplished in the "real world." Mentoring is not necessarily a friendship. You should not anticipate extending your relationship past the workplace. If you make a connection with a coworker that is a comfortable working partnership, adopt him or her as a mentor. Ask the person to teach you and give you opportunities. Pick his or her brain. Use staff members with more experience to demonstrate their expertise. Devoted veterinary professionals want proper care for their patients, and everyone benefits if you learn to do your job to the best of your ability. Mentoring develops leadership and consistency. If you learn a procedure from a mentor, hospital protocol is consistent and fewer conflicts arise. In time, you may be surprised when the newest employee approaches you for help.

We have discussed many factors that contribute to the development of good people skills. Have you recognized traits in yourself that you want to modify or improve? That's great, because when it comes to dealing with people, all of us have areas that can use improvement. While you are still in an academic environment, take time to practice your people skills. Every day, speak to someone you do not know. Say "hi," make eye contact, and smile. Accept the challenge of trying to get someone who appears in a sullen mood to speak to you and smile. Make it your personal challenge. Some people do not find this a difficult task, but others find it nearly impossible to initiate a conversation or meet someone's gaze. A little practice now will help you make a smoother transition into the workplace.

CONFLICT RESOLUTION

The Angry Client

Handling people who are upset or angry takes practice and a plan. Angry clients can cause considerable turmoil in a veterinary practice, and most veterinary personnel are happy to hand off this particular problem to the office manager or the doctor. There will be occasions when you, either in person or over the phone, will have to deal with an angry client. When you encounter this situation, you can use the people skills you have learned to effectively diffuse the conflict and take control. You should never tolerate abusive behavior, but you should not mirror it back either. Learning to control a confrontational situation is a little scary. Most people would rather avoid confrontation, but a few steps will help you survive it.

1. Determine why the client is angry.
2. Let him or her talk; don't interrupt.
3. Maintain eye contact, and nod your head so the client knows that you are following his or her words. If your client is on the phone, make appropriate acknowledgements that you are listening, but do not interrupt.
4. Paraphrase. Make sure you understand the problem.
5. If you can offer service, do so. If not, give the client a time frame for resolving the problem and following up.
6. If you feel physically threatened at any time, do not try to resolve the problem by yourself.

Most client confrontations come from anger and frustration due to a breakdown in communication. It is essential that you keep each and every client informed about his or her pet's diagnosis, its condition and prognosis, and the financial expectations for its care. While most of these conversations will occur between the client and the veterinarian, rest assured that the client is going to ask you questions. Also be aware that you can strive for and deliver flawless customer service and still have clients who are angry. You should expect this. Every practice experiences these confrontations.

The Angry Coworker

It is also common for conflicts to arise between coworkers. Sadly, many of these battles go unresolved or, even worse, become feuds. Employees cannot serve the best interests of the patients and the hospital if they have internal conflicts. If there are personality differences, disagreements, resentments, or misunderstandings, you must put these issues to rest. Unsettled issues lead to poor job performance and can result in the decision to leave your otherwise-satisfying employment. No one wants to come to work and "walk on eggshells" around another person. A strategy for ending a conflict with a coworker should start with an evaluation of the problem. Is this a petty problem that will go away? If your coworker failed to replace the roll of bathroom tissue in the break room, it may not be worth the effort to confront him or her. If there is a more serious issue that needs attention, consider the following:

1. Is the problem personal or work related?
2. Determine whether the problem needs input from your supervisor. If the conflict involves a breach of professional ethics, the law, or a matter of confidentiality, ask the appropriate people to sit in on the conversation. Examples include falsifying medical records, altering controlled drug logs, theft of property, lying to clients, and performing treatments without consent.
3. Approach your coworker. Ask for his or her attention. "Tammy, may I talk to you a minute?"
4. State the problem. "I've noticed that you've left early the past several weeks, leaving the rest of us to close the hospital. We've been here late cleaning up every time this happens."
5. Ask for feedback. Tammy might say, "I'm really sorry, but I have to pick up my kid at the sitter before 6:00 p.m."
6. Suggest that you work together as a team for a resolution. "What can we do to balance the work the others are doing in your absence?"
7. If you feel physically threatened, do not try to resolve the conflict by yourself.

Part of maturity and professional growth is to resolve conflicts, accept the solution, and move forward without resentment or retaliation. Too much mental and emotional energy invested in a petty conflict can be exhausting, and detracts from the important job of caring for patients. Actions by coworkers that negatively impact your job performance must be resolved. Never "cover" for illegal, unprofessional, or unethical behavior. Failing to resolve such matters can implicate you as a participant.

MEETING YOUR EMPLOYER'S EXPECTATIONS

You may be required to go through a long process before you land your desired job. Interviews, background checks, references, and crafting a résumé are all part of what you can expect when job hunting. By the same token, your employer goes through a long process to find the person who best fits the clinic's needs. There is a significant investment in time and money to write and place an advertisement for a position, interview qualified candidates, pay for background checks, and verify references. Once you have secured the job, there is another investment in training time, salary, benefits, and getting you to be a competent, independent worker. Your employer is hoping that not only will you live up to the expectations that you put on your résumé, but that you will stay at your job for a significant amount of time. Here are some tips to help ensure success once you land your job.

1. Punctuality: Your employer expects you to be on time every day and ready to work.
2. Dependability and giving an honest day's work: Give a full day's work for your salary. Do not run personal errands on company time. Do not "pad" the time clock by staying longer than necessary.
3. Teamwork: Work as a team member. Support the hospital's needs and goals versus personal ones.

4. Appearance: Your appearance should be neat and clean. You are representing your team!

5. Loyalty to your organization: Promote your clinic; be a walking spokesperson.

6. Knowledge of operating procedures: You should know your clinic's policies and procedures, and know the location of vital information.

7. Knowledge of materials and equipment: Hospitals make significant investments in equipment. Your employer expects that you can properly use and maintain diagnostic and support equipment and materials.

8. Working without close supervision and following instructions: Be a self-starter. You do not need someone to babysit you!

9. Working under pressure, meeting deadlines: You will be under significant pressure most days to care for patients and meet other work demands. Work efficiently and be aware of deadlines.

10. Managing time and materials effectively: Use your time and clinic resources effectively. Conserve supplies when possible.

11. Safety regulations: Follow safety regulations—always! There is no compromise here. Many of these regulations are federal laws. They are designed to protect you and decrease the liabilities of the practice.

Employers are often as interested in your work ethic as they are in your education. Technical skills improve with time and practice, but you will not stay at your job long enough if you are continually late or slack your duties at work. Look inside and ask yourself, "Am I someone who I would hire to work for me? Would I be satisfied with my level of professional behavior?" If you see areas for improvement, make a list and tackle them one by one. Engage your coworkers to help you. Ask family and friends for support. Look for the solutions that allow you to dedicate 100 percent of your head and heart to the patients depending on you to do your very best.

SUMMARY

You are engaged in a commitment to animal health and welfare and are beginning to grasp the complexities of providing good care by understanding animal anatomy, physiology, and behavior. You must also understand and relate to the behavior of clients, coworkers, and employers. There are expectations in the workplace that require your best appearance, cooperation, and teamwork. You always want to put forth your best effort and represent yourself as the kind of person that *you* would trust to take care of *your own* animal. It is a worthwhile investment to spend time polishing your "soft skills."

Soft skills, also known as "people skills," can be a challenge for the veterinary technician student. Many of you have entered the field with the expectation that working with animals relieves you of having to deal with people. In reality, working with people becomes vitally important, and you must devote some time and practice to developing the skills that make people feel comfortable and at ease with you. You have to instill enough confidence in your client to allow him or her to entrust to you the care of a pet.

Nonverbal communication is a large component of your intent and feelings. Your body language and eye contact make this apparent. "Mirroring" is a technique that we use to put people at ease by mimicking positive behavioral postures. It is a signal that says, "Relax, I am like you."

Verbal communication requires thoughtful listening to and acknowledgment of the speaker, showing that you are engaged and interested in the information that he or she is sharing. This involves signaling with positive gestures such as nodding or cocking the head, verbal assurances such as "uh-huh," and paraphrasing the message back to the speaker.

Your professionalism is reflected in your interactions with clients and coworkers. Positive professional habits include proper attire and attitude, keeping the work space clean and organized, and investing in your professional growth. You may wish to adopt a mentor to increase your knowledge and confidence as you progress through the early stages of your career. You must carry out conflict resolution with both clients and coworkers in a dignified and organized manner. Make sure that you are calm and that you diffuse tempers before engaging in any discussions. Typically, angry people just want to be heard and given a time frame for problem resolution. You must terminate any conflict that results in the threat of physical harm, by removing yourself from the area.

Be an asset to your employer. Adopt the attitude of how valuable your job is to you, not what your job can do for you. Be punctual, dependable, and flexible, and also live within the policies and procedures your practice sets.

TEST YOUR KNOWLEDGE

1. Nonverbal communication is an important component of the messages that we convey to one another. Examine the next three photos in which only the eyes are visible. Explain what aspects you observe and what message each photo conveys.

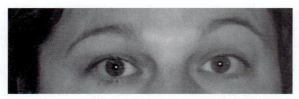

Amy Wolff

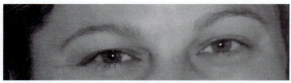

Amy Wolff

Amy Wolff

2. Describe how you would use nonverbal communication to make a new client feel relaxed and at ease during his or her first clinic visit.
3. With regard to safety, explain the importance of professional dress in the workplace.
4. Describe three characteristics of professional behavior in the workplace
5. Explain the advantage of maintaining a neat and organized work space.
6. Discuss the benefits of adopting a mentor. What would be your expectation in developing a mentor relationship with a coworker?
7. Give three examples of workplace behavior that are illegal or unethical. Describe your course of action for reporting these events.
8. Create a plan to make "deposits" in your emotional bank. Identify activities that restore your energy and reduce stress.
9. Discuss the process of conflict resolution with an angry client. Describe the proper action for responding to a situation in which you feel physically threatened.
10. Summarize employer expectations for workplace performance. Identify personal obstacles that might keep you from fulfilling these expectations, and formulate solutions to allow you to meet your employer's requirements.

BIBLIOGRAPHY

Bayer Animal Health Communication Project. 2012. "Home page" Last accessed July 31, 2012 from the Bayer Animal Health Communication Project website.

6

The Medical Record

Chapter Outline

Learning Objectives

At the end of the chapter, you should be able to:

- Describe the reasons for keeping medical records.
- Explain the legal requirement for the retention of medical records.
- List and describe the documents that belong in a medical record.
- Discuss the ownership of medical records and the release of information.
- Compare the advantages and disadvantages of paper versus electronic medical records.
- Explain the differences between the SOMR (source-oriented medical record) and the POMR (problem-oriented medical record).
- Utilize findings from a physical exam to write a case history in SOAP form.

INTRODUCTION

The veterinary hospital is a busy place. Exam Room 1 has a Labrador Retriever that has been vomiting for three days, Exam Room 2 has a cat with hair loss, and Exam Room 3 has a bird that needs its wings and nails trimmed. To keep the workload flowing smoothly, there is a division of labor. Only a licensed veterinarian will form a diagnosis, prescribe medication, and perform surgery. The practice manager's duties include scheduling, payroll, accounts receivable, and the day-to-day business

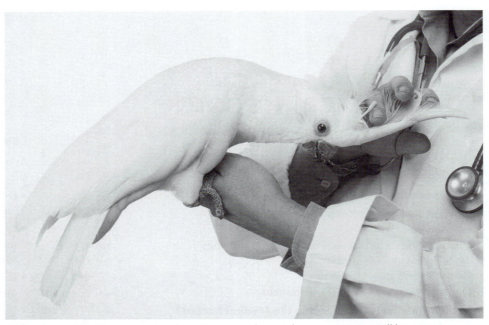

A veterinary clinic is an amazing place. You never know who your patients will be.
Roger Costa Morera / Shutterstock

of running the clinic. The technician's role includes patient treatments and assisting the doctor in the various diagnostic and treatment protocols. However, there is one task that involves all hospital personnel: generating and maintaining accurate medical records. With all these different activities occurring in the clinic, the medical record is the way that the veterinary team communicates with each other regarding their part in patient care.

You may be familiar with medical records as they relate to your own health care experience. Your health care providers serve many patients, and it is impossible for them to recall the important details of every patient by memory alone. The health care provider creates a permanent record so that it can organize your health history in an accurate and accessible way. In this chapter, we will discuss the importance of creating and maintaining veterinary records with the same care and accuracy as our human health care counterparts do. At the time a patient first enters the veterinary clinic, a staff member will start a medical record that will provide documentation and descriptions of the patient's health. The record will contain many types of information, and any hospital staff member who is involved in patient care should maintain and update it.

The medical record should be a complete description of the patient's experiences in the hospital for both wellness and illness. As an essential part of the veterinary

You would not be able to recall all the medical information about these dogs without the help of accurate medical records.
Erik Lam / Shutterstock

team, you will make frequent entries into a patient's medical record. For that reason, we will explore the purpose of the medical record, types of information included in a medical record, why accurate record keeping is critical for the patient and the hospital staff, the legal aspects of record keeping, and writing good basic notes that clearly summarize and detail the patient's care. Lastly, we will explore some of the many options for storing and organizing patient records. Remember that a medical record can include information of varied types, including printouts, notes, log entries, radiographs, and CDs. It is important to integrate and store these components in a manner that supports patient care.

THE PURPOSE OF THE MEDICAL RECORD

The primary purpose of the medical record is to support patient care. It serves as a history of every visit and illness, as well as of routine health. Over time, it may illustrate a trend that helps identify a disease process, or simply document changes that accompany the aging process. The patient's record also includes procedures, prescriptions, lab test results, and phone conversations between hospital personnel and owner. The medical record also contains information about the patient's owner(s), their contact information, and any special instructions they wish the hospital staff to know. A record is vitally important, as most of us cannot remember the details of every patient's visit or medical care in the hospital. You will not be able to remember the name, dose, and route of a drug that your hospital dispensed a year ago. It is the best way for every staff member to have information available as different personnel within the clinic assume the duties of caring for that animal. *Good medical records support good patient care.*

It is important to keep a separate and complete medical record for every patient. If a client has several pets, keep the records in a single folder, but distinguish each pet with its own record. An exception to this is in food animal/large animal medicine. Records there usually refer to the herd. Typically, all herd members have the same medical and surgical procedures, as well as vaccines at the same time. It is not realistically possible to keep a record for each animal unless one has an individual problem that the veterinarian must treat.

Practitioners treating equine and food animal patients must be mobile. It is impractical to keep extensively detailed records on flocks and herds in a vehicle. So much of what the veterinarian records is on a standardized form with attached duplicates that he or she can give to the owner and any state agencies. Notebook or laptop computers are portable, compact, and help practitioners stay organized. It is less likely that the veterinarian will overlook treatments and charges if prompted by veterinary software. The veterinarian can easily download information to a central system or send it electronically to the hospital. Equine medical records are more centered on individual patients. Horses that are companion animals will need records that indicate their vaccination histories, document parasite control, and give a description of any health or structural problems. If an owner transports a companion horse for shows or competition, updated records and health certificates showing vaccination status and Coggins testing for equine infectious anemia must be current.

IMPORTANT REASONS TO KEEP MEDICAL RECORDS

Personalized Care

You can also use medical records to **personalize** the care that you give your clients. Although this is not directly related to your patients' health, you can make a note of important facts about your clients so that you can make casual inquiries when they

A laptop makes it easy to stay organized and keep accurate records.
Goodluz / Shutterstock

visit your practice. Making a note of the names of children in the family, birthdays, or special events can remind you to ask the client a personal but neutral question. "Hi, Mrs. Jones. How is Julie doing now that she's away at college?" What a great way to impress Mrs. Jones! Also make notes that are helpful for personal information but are not actually topics of discussion. For example, some hospitals will make a note in the record if a spouse, partner, or child is ill or has passed away. Although you do not discuss these issues with your clients, it may keep you from making an embarrassing mistake. The greeting, "Hi, Mrs. Jones. How are you? How's Mr. Jones?" could cause a tense or hurtful moment if Mr. Jones is deceased. This also applies to pets. Many of your clients will have multipet households. If a pet has passed away (especially if it was euthanized in your hospital), it makes a poor impression to inquire about its health. You can also make brief notes in the record about problem behaviors. Many pets are fearful and anxious when they visit the clinic. You may need to restrain or muzzle them for certain procedures. Some do better when you take them from their owners. Making a note in the chart explaining these situations keeps your coworkers from getting hurt and makes for smoother visits. Do not forget the small details. Make notations if you are to address the client in a specific manner. There are generational and professional expectations in this regard. Despite the trend to greet everyone by their first name, many people by convention prefer to be addressed as Mr., Mrs., or Doctor. Another idea is the use of colored paper for coding the record—e.g., using pink paper for female patients and blue paper for male. It may sound silly, but clients really dislike it when you use the wrong pronoun when addressing the pet. The color codes help you avoid that mistake.

Charting Trends in Diseases or Specific Conditions

You can use patient records to chart trends in diseases or conditions. In the case of infectious or zoonotic diseases, veterinarians may need to provide information to state agencies regarding the types of conditions they see in the hospital, as well as the number of affected animals. This is particularly important in herd health management and protecting our food supply.

Occasionally, colleges of veterinary medicine will send letters to veterinary practitioners asking if they would like to participate in clinical trials that involve specific conditions. You can see from the Sample Trial Letter on page 176 that the University of Missouri College of Veterinary Medicine, Department of Ophthalmology, is asking for veterinarians to refer cases of keratoconjunctivitis sicca (dry eye) for specialized studies. If a clinic has kept accurate medical records, personnel can identify this patient population, contact the owners, and refer the pets for specialized treatments. Medical records can also provide research documents and statistical analysis. Let's imagine that you want to write an article for the National Association for Veterinary Technicians in America (NAVTA) journal on your hospital treatment protocol for canine Parvovirus. You will need to identify the number of cases that your clinic has seen, what treatments it used, and how the patients responded. It would be a very difficult task to recall all that information from memory. That is why the medical record is such a vital tool. Properly maintained, it provides all the necessary information regarding these patients and how the hospital or clinic managed the cases. Because this type of information is so important in recognizing disease patterns and contributing research data, the Veterinary Medical Database was established by the National Cancer Institute in 1964. Originally, the purpose was to track the occurrence of cancer in companion animals, but it now compiles information from twenty-six schools of veterinary medicine into a searchable database (Veterinary Medical Data Base, 2010).

Tracking Financial Information

The medical record also retains information regarding clients' financial accounts with the hospital and should contain information regarding balances, accounts receivable, credits, and any discounts applied. Some clinics offer discounted services for senior citizens, military personnel, or clients who refer a large number of new clients. If a client is entitled to one of these discounts, it is important to make a notation to ensure smooth service at checkout. Staff should also make notations about nonpayment of accounts, to protect a hospital from providing services to a client who has not met previous financial obligations.

Let's say that Mr. Harrow comes in to All Pets Animal Hospital wanting vaccinations and a nail trim for his dog Peppie. A quick check of the record indicates that Mr. Harrow owes the clinic $200.00 from an emergency visit when Peppie had an episode of pancreatitis. Because the record is clear and up to date, staff members can be aware of the issue and alert the doctor or practice manager before they render more services.

Documenting Conversations

Over the course of a patient's evaluation, there may be numerous conversations with the owner about diagnosis, treatment options, prognosis, and fee assessment. The owner may also seek second opinions from other veterinary specialists and discuss this with you. Documenting these conversations and any decisions the owner makes regarding patient care is important, to protect the hospital and staff from liability.

On a busy practice day, a client comes in with her three Jack Russell terriers—Rascal, Max, and Lacey—for their annual checkups and vaccinations. They have boundless energy and are making a big tangle of the flexible leashes that Mrs. Wright is holding while they lunge for the clinic cats. You are in the back preparing the vaccines, and your coworker is weighing the dogs and preparing the check-in sheets. One at a time, the doctor conducts the physical exams and administers the vaccines while you restrain the dogs. As the doctor presents the physical exam findings for each patient, he looks down to see that Lacey is now covered in hives and is experiencing severe facial swelling. Mrs. Wright is very upset. Lacey had this happen last

Sample Trial Letter

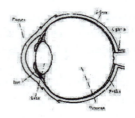

The Ophthalmology Service at the Veterinary Medical Teaching Hospital at University of Missouri-Columbia would like to announce a clinical trial for a promising new therapeutic agent for the treatment of dry eye (keratoconjunctivitis sicca or KCS) in dogs. We are also offering $100 to your practice for each successfully enrolled dog.

We are currently seeking newly diagnosed, <u>untreated</u> dogs over 1 year of age, of any breed or sex to enroll in a treatment trial. Dogs should have a diagnosis of idiopathic KCS (presumed immune-mediated). Dogs that have suspected pharmacologic, toxic, infectious or traumatic causes are not eligible for inclusion in the study. <u>Dogs *must* be naïve to cyclosporine, tacrolimus and steroids but treatment with other medications and artificial tears/ointments is acceptable.</u>

Owners will receive free ophthalmology examinations for the 3 month duration of the trial, as well as free trial drug during that time period. Owners will also receive free cyclosporine for 12 months after their pets complete the trial. The cyclosporine alone is approximately a $460 value for the owners, in addition to the free examinations for the first three months. Owners will be required to bring their dogs in for monthly rechecks at the VMTH. At the completion of the trial the patient will be referred back to the referring veterinarian for ongoing management.

Referring veterinarians/clients may also call to speak with a member of our ophthalmology service directly at   for more information.

Rascal, Max, and Lacey are presented for physical exams and vaccinations.
Erik Lam / Shutterstock

time, and Mrs. Wright called the practice to report the reaction. She says that the staff member provided her with instructions on what to do for Lacey, and she was assured that the doctor would give the dog medication to decrease the chances of a vaccine reaction before the next injections. A thorough search of the medical record fails to turn up any notations about Lacey having vaccination reactions or of Mrs. Wright's phone call.

This is a simple but important example of why it is critical to ensure that you log *all* patient information into the medical record. The person who spoke to Mrs. Wright should have pulled the record immediately and recorded the incident, as well as any of the clinic recommendations. This person could highlight the conversation in the record so it would attract instant attention, and then the individual handling the situation could take appropriate steps prior to vaccination. The lesson learned from this example is clear: make proper notations in the medical record as soon as possible. A common legal analysis of a medical record in question is "If it isn't written down, it didn't happen." Make excellent record keeping a priority.

Documentation Is a Legal Requirement

Lastly, the medical record is a legal document, and you should always approach it with the thought that every remark or notation you make becomes part of a legal record. For this reason alone, you want to make clear, concise, professional comments in your records, free of emotion or personal editorializing. No matter what the situation, do not make any disparaging remarks about the patients or the clients. Before you make any notations, pretend that a judge in a court of law will read your entries. How would you want them to sound? As tempting as it may be to doodle, draw, jot reminders, or perform calculations in the margins, only use the medical record to document the patient's care. Records should be neat and legible and kept in black ink. Do not use brightly colored pens or writing styles that are unprofessional. Initial and date every entry. It is also necessary to note the time of day, especially if you are having multiple conversations with a client. Equally important is the manner in which you maintain the record. Do not manipulate or alter it in any way. You must never erase, scratch out, or white out any information in the record. Others can perceive it as trying to hide information or errors. If you do make a mistake in the record, draw through it with a single line. Write "error" close to the text, and initial the change. That particular mistake happens to all of us, and there is no penalty for it. Make a careful attempt to correct the misinformation, and then insert the correct data. Sometimes you will write down a long patient note or client conversation, only to find that you have entered it in the wrong record! It is frustrating, but correct the error and reenter it in the proper patient's file.

We have emphasized several times that it is important to record even the smallest details in the medical record. There is a good reason for this. Oftentimes, a client will bring up an issue to the veterinary hospital months to years after the event. In the meantime, he or she may have thought about an unfavorable outcome, done online research, or networked with others who may influence his or her decision to pursue litigation over a case that your hospital assumed was worked to its conclusion. When that happens, the hospital will be able to produce evidence in the form of notes and documentation of conversations, dated and initialed by the staff, regarding the pet's condition.

Keeping an accurate medical record is not just sound medical care—it is also a requirement of the practice acts of state veterinary medical boards or other licensing agencies. The individual state determines how long a clinic or hospital must maintain medical records. A period of five to seven years is typical. For ease of organization, and to maximize space, you can put inactive records in storage. After the statute has expired, you can delete or destroy records permanently. For security purposes, never throw these records in the trash intact, leaving personal information vulnerable. Your clients are better protected if you use a shredder.

For review, the list below summarizes the reasons why we keep medical records:

1. It is the best way to document a patient's health status and provides consistent communication among the veterinary team members, to insure that everyone has access to the information necessary to provide the best possible care.
2. Medical records also carry information about the owner, how he or she wishes to be contacted, and his or her account with the veterinary hospital.
3. We can use medical records to compile information for research projects or specialized studies.
4. We can use medical records to analyze patient data and treatment protocols, to determine response to therapy.
5. Medical records can personalize client visits.
6. It is a legal requirement to keep a medical record for every patient that the veterinary team sees, even if the problem seems insignificant.

Case Scenario:
"All Pets Animal Hospital. This is Holly. How may I help you?"

"Hi, Holly. This is Jerry Morris, and I have Ruffy, who woke me up today whining to get outside. He had a bad case of diarrhea like he did a month ago. He keeps getting in the trash. I just don't know how to break him of this habit. I gave him some Pepto-Bismol™, and he's fine now. Anyway, I just really called to refill his thyroid medication. Can I pick up the medication around 5:30?"

"Sure, Mr. Morris. I'm glad Ruffy is feeling better. See you tonight."

As Holly prepares the medication, she notes the refill order but makes no mention of the diarrhea or the fact that Mr. Morris gave Ruffy some medication. Because Ruffy's diarrhea has stopped, she thinks there was really no reason to bring it up.

Is Holly's omission a critical one? Yes. Even though Ruffy had only one episode of diarrhea, which has now resolved, the owner communicated a medical issue with his dog and gave his dog medication that could have had side effects. Holly should have made an immediate notation in the record that included the date and time, with whom she spoke (Mr. Morris), and the pertinent context of the conversation. The veterinarian should have then determined whether any follow-up was necessary. This information may become important sometime in the future, even months later. Document the medical record every time a client relays any information to you about a patient—whether it is in person, over the phone, or electronically through fax, e-mail, or online conferences.

WHAT INFORMATION IS CONTAINED IN THE MEDICAL RECORD?

In a word, everything! Let's take a look at the list below to become familiar with the components of a typical file. The veterinary technician should have a complete understanding of what goes in a medical record and why. Even though you will not be responsible for the patient's diagnosis, you will be responsible for ensuring that you include the proper data in the record.

1. Client information. The medical record data begins with client information. Most hospitals will use a standardized form with their hospital name and logo. Minimally, client information includes name, address, and phone numbers. Data that provides more-detailed information would include e-mail address, how and when the owner would like to be contacted, and an emergency contact if the client is not available. It should indicate payment preferences. If the client has a religious, ethical, or moral issue that he or she wishes to include as a guide to patient care, note it in this information sheet. Some hospitals require state identification or Social Security number if the client wishes to write a check as payment.

2. Patient information. It is important to have accurate, organized patient information. This is not always available with veterinary patients, so you must do the best job you can with what the client can offer. Patient information begins with the **signalment** and **history.** The signalment consists of three pieces of data: the breed of the animal, its age, and its sex (reproductive status is also indicated). This is usually on the first page of the record or the file cover (see the Sample Client/Patient Information Sheet on pages 180–181). Think of the signalment as the first bit of information that you want to know about your patient. Why is the signalment so important? It is because it helps the veterinary team focus on the most common diseases and conditions for specific patients. For example, imagine that you are in the back of the clinic, when you hear the receptionist talking on the phone. She says a client is bringing in her dog 'Cassie' for vomiting. At this point, we don't know anything except a dog is coming in for vomiting. That leaves us with many open questions, so we ask for more information. What kind of dog? How old is it? The receptionist tells us that it is a female (intact) Golden retriever puppy that is twelve weeks old. Ah ha! Now we have something to go on. Try to name three medical conditions that might cause vomiting in a dog this age. Let's change the scenario. What if the receptionist told us it was a *twelve-year-old* female intact Golden retriever? What medical conditions might cause these symptoms in a dog of this age? The signalment helps us formulate ideas about the potential causes for an animal's illness.

The animal's breed is also significant. In the example, the dog coming in for vomiting is a Golden retriever. This breed has a tendency to consume foreign objects, especially as young dogs. (Did you include the consumption of a foreign object in your list of possibilities for the vomiting puppy?)

Sample Client/Patient Information Sheet

All Pets Animal Hospital

"Dedicated to service, devoted to care"

Patient/Client Information

Welcome to All Pets Animal Hospital. Thank you for giving us the opportunity to serve the health needs of your pet. Please help us by completing the information sheet so we can provide the best care for your pet and your family.

Name & Title _____ Spouse/Partner/Other _____

Address _____ City _____ Zip _____

Contact Phone Number _____ E-mail _____

OK to leave message on voice mail? Yes/No   Text message? Yes/No (please indicate)

Employer _____ Employer Telephone _____

In Case of Emergency, please call _____@ Telephone _____

Preferred method of communication? Phone/e-mail/text (please indicate)

How did you select our hospital?
Hospital Sign   Convenient   Yellow Pages   Newspaper
Website   Other   Referred by _____

Professional fees are due at the time services are rendered. We will gladly provide you with a written estimate for services. A 50% deposit is required for all pets admitted for hospital stay, with the balance due at the time of discharge. We accept cash, debit cards, and VISA, MASTERCARD, DISCOVER. We will accept a check from a local bank with a valid driver's license. There is a $50.00 fee for returned checks.

To prevent the spread of infectious disease and parasites, we recommend a program of vaccination and parasite control. Pets with fleas will receive topical or oral medications on admission, and the cost will be included on the invoice at the time of checkout.

Patient Information

Please tell us about your pet!

Name _____ Species: Dog Cat Bird Reptile Other _____ Breed _____

Approximate Age _____ DOB _____ Sex _____ Neutered/Spayed/Intact

Color/Markings _____ Is your pet microchipped? _____

Sample Client/Patient Information Sheet continued . . .

Vaccination History (Please indicate the last time your pet was vaccinated for the following.)

Dogs:

Rabies _____ Distemper _____ Bordetella _____Lyme Disease _____ Canine Influenza

Cats:

Rabies _____ Distemper/Upper respiratory _____ Feline Leukemia _____

Tell us about your pet's lifestyle:

Activity level: High _____ Medium _____ Low _____

Exercise: 1–3 x per week _____ 3–5x per week _____ More than 5 x _____

Diet: Commercial (please indicate brand) _____ Prepared (please describe) _____

Does your pet have any known allergies or sensitivities? Yes/No (please indicate)

Is your pet taking a heartworm preventative? Yes/No   If yes, what product? _____

Are you using a parasite control program for fleas? Yes/No   If yes, what product? _____

What health concerns do you have for your pet?

Please describe _____

_____

_____

_____

Please describe _____

_____

_____

_____

Do you have behavior issues you wish to discuss?

Please describe _____

_____

_____

_____

As you gain experience, you will begin to see patterns emerge that are tied to the signalment. You should begin every conversation about a case with the signalment. It should sound like this: **Cassie is a 12-week-old female intact Golden retriever** that has had three episodes of vomiting in eight hours. (The signalment is the portion indicated in **bold** print.) When you come to your veterinarian with such concise information, it starts those mental wheels turning in the right direction. After establishing the signalment, the process of gathering pertinent information about the patient's health begins. We refer to this information as the *history*. This will include a wide range of questions. You will ask the owner to provide information pertaining to all aspects of the pet's life. Vaccinations, lifestyle, nutrition, exercise, and behavior are all part of a comprehensive history. This collection of information should include past and current events. The history is one of the most important pieces of information in the record. You can often see patterns that can help lead the veterinarian to the diagnosis. Before you enter an exam room or engage in a conversation regarding patient care, take out the patient's record and read the history. Look at the Sample Single Animal Record on pages 183–184 for Sassy, a six-year-old Lhasa apso F(S) who has come in for itching. See if you can find a pattern in the history that helps guide you to the most probable cause of her pruritus.

The patient history is an important tool, but we still must approach every patient with an open mind. To rely too heavily on history is to miss the possibility that something else is happening.

3. Physical exam reports. The physical exam report is the description of the patient and its clinical signs as observed by the veterinary team (see Sample Exam Report Card on page 185). It may include comments from multiple staff members. It often starts with your entry of the patient's weight and vital signs. Much like when you visit your own health care provider, your physical exam starts with an assessment of your height, weight, and blood pressure and entering the values in your chart. Some hospitals will use a preprinted form to record the findings of the physical exam, and others will keep handwritten notes. It is crucial to record the patient's examination findings. Not only will it serve as legal evidence that an examination of the problem has taken place, but it also serves as an account of what you saw at the time of the examination. Look at the sample record for Duke, a three-year-old quarter horse gelding that requires an examination for a wound that he sustained while trying to go through a barbed-wire fence (see Sample Single Animal Record on page 186). Note how the doctor indicates recommendations and observations in the notes as she visits the patient. Why do you think this is helpful for the doctor and the clinic?

4. Laboratory reports. Almost every patient that you see will require some type of laboratory work. Healthy patients coming in for routine care frequently have fecal samples, heartworm tests, ear swab cytology, skin scrapes, or aspirates to evaluate. Patients that the veterinarian evaluates for problems related to illness may require other more specific lab tests such as a complete blood count (CBC), serum biochemistry profile, urinalysis, or tests for endocrine function. You should make printed or handwritten notes of the results in the patient record (see Sample Blood Results Form on page 187).

5. Anesthetic reports. The anesthetic report is a detailed record of events from the preanesthetic physical to complete recovery. It will include entries for the types of drugs that the patient received, the method by which they were administered, the patient's vital signs while under anesthesia, oxygen and carbon dioxide levels, fluid administration, and any other supportive measures. (see Sample Anesthesia Monitoring Form on page 188).

6. Surgical reports. The veterinarian writes a detailed report following a surgical procedure. For common procedures with low complication rates, such as spays and neuters, many clinics use a template to save time (see Sample Surgical Report Form on page 190). The veterinarian may enter additional notes describing any

Single Animal Record

Client Name: Quick, Joe **Address:** 999 Locust Street, Anywhere, USA **Phone:** 555-2222

Pet Name: Sassy   **Species:** K9   **Breed:** Lhasa   **Sex:** F(S)   Age: 6 years
DOB: 12/22/2005   **Color:** B/W

MO	DAY	YR		Temp	MM	CRT
1	17	2005	Presented for first vet exam. Obtained from breeder last night. No immunizations/wormings to date. Puppy chow diet. Normal physical exam. Fecal exam- roundworms/hookworms. Wt. 4.2 lbs. Tx- Panacur 1/4 tsp PO QD x 3 days DHPPC SQ Recheck 3 weeks	100.0°F	Pink	1 Sec
2	7	2005	Presented for vaccination, recheck fecal. Puppy doing well, owners having trouble housebreaking. Appetite good, no V/D. Normal PE. Fecal exam- Neg. for ova Wt. 6.8 lbs. DHPPC SQ Repeat Panacur 1/4 tsp PO QD x 3 days. Recheck 3 weeks.	100.0°F	Pink	1 Sec
2	28	2005	Presented for vaccination. Housebreaking improving. Doing well. Normal PE. Recheck 3 weeks 9.2 lbs DHPPC SQ			
3	21	2005	Presented for vaccination. No problems reported. PE WNL. Wt. 11 lbs. Rabies 1 yr SQ (tag # 11188) DHPPC SQ.			
5	15	2005	Admit for spay. 10.8 lbs. See surgical report. Wt. 12 lbs.			
5	25	2005	Medical progress exam- Doing well. Sutures removed.			
9	12	2005	Presented for pruritus, 2 weeks' duration. Rubbing face, chewing feet. Generalized dermatitis, no lesions. Skin scrape neg. DTM- neg. Rx- Hydroxyzine 10 mg #14 1 tab PO BID x 7 days Wt. 13 lbs.			
9	21	2005	Owner called, dog improved, wants refill on meds. Rx- Hydroxyzine 10 mg #14 1 tab PO BID x 14 days			
3	4	2006	Presented for yearly vaccinations. No problems reported. PE- WNL. DHPP, Rabies 1 yr (tag # 19311) Fecal- neg. Wt. 18 lbs. HW- neg. Rec. monthly heartworm and flea control.			
10	13	2006	Presented for itching, 2 days' duration. Dog is rubbing face, scooting, licking feet. Generalized dermatitis and secondary staph. Skin scrape- negative. DTM- neg Rx- Cephalexin 250 mg #42- 1 cap TID x 14 days. Hydroxyzine 25 mg #14- 1 tab PO BID x 7 days			

Single Animal Record continued . . .

	Date				Temp	MM	CRT
3	11	2007	Presented for vaccination. Owner reports foul breath. Also seems lame on the right rear leg. PE- alert, eyes- OK, ears- OK, oral cavity- grade 3 dental tartar, lymph nodes- OK, hydration- OK, H/L- OK, ABD- OK, musculoskeletal- obese, BCS 8/9, no signs lameness, no pain elicited, no heat or swelling. Integument- OK, Urogenital-OK. Rabies 1 yr (tag # 12545) DHPP Rec. dental cleaning ASAP. Radiographs RR leg while under sedation.			
3	15	2007	No show for dental. Called 8:00am- left message.			
4	24	2007	Presented for bite wound to the dorsal cervical area and right front paw. Dog in fight with stray that entered yard. 1 puncture wound on the dorsal aspect of the neck approximately 2" caudal to the right pinna. 1 puncture to the anterior aspect of the right front paw. Clip, clean, and flush with chlorhexidine soln. Clavamox 250 mg #14- 1 tab PO BID x 14 days. Carprofen 25 mg #7- 1 tab PO QD. Wt. 21 lbs. Progress call in 3 days.	101.5°F		
4	27	2007	Progress call 10:30am- dog doing well.			
5	31	2007	Presented for diarrhea, 24 hours' duration. Dog was given a basted rawhide bone yesterday, consumed about half the bone. 2 episodes vomiting. Tx- Cerenia 1.5 ml SQ. NPO 12 hrs, then bland diet x 3–4 meals. Rx- metronidazole 250 mg #14- 1 tab PO BID x 7 days			
6	4	2007	Progress call- diarrhea resolved. Told owner to finish all meds.			
9	15	2007	Presented for pruritus, 4 days' duration. Dog has been keeping owners up at night scratching, now had sores on face, paws, and one by the tail. Generalized dermatitis with pustules and epidermal collerettes. Skin scrape- cocci bacteria, yeast. Wt. 22.5 lbs. Rx- Cephalexin 250 mg #42- 1 cap PO TID x 14 days Rx- Ketoconazole 200 mg #14- 1 tab PO BID x 7 days			

Sample Exam Report Card

All Pets Animal Hospital
Examination Report Card

Pet's Name _____   Owner _____
Reason for Visit _____   Date _____

Contagious Diseases Prevention Program

_____ Up to Date
__ Vaccination due: Distemper __ Rabies __ Bordetella __ Influenza __ Lyme __ Leukemia __
___ Vaccinations Administered: Distemper __ Rabies __ Bordetella __ Influenza __ LymeLeukemia __
_____ Annual: Heartworm test __ Fecal exam __ Flea/tick control __

Systems Review

1. Integument/Coat Condition

Appears normal ___
Dry/flaky ___
Unkempt/matted ___
External parasites ___ Type_____
Hair loss ___ Pattern _____
Wounds/lesions ___
Mass/cyst/tumor _____
Pruritus _____
Other _____

2. Eyes

Appears normal___
Lenticular sclerosis _____ O.S/O.D/O.U
Cataract _____ O.S/O.D/ O.U
Conjunctivitis ___ O.S/O.D/O.U
Discharge _____ O.S/O.D/O.U
Ulcer _____ O.S/O.D/O.U
Other _____

3. Ears

Appears normal ____
Discharge ____ A.S/A.D/A.U
Hematoma ____ A.S/A.D/A.U
Mites ____ A.S/A.D/A.U
Mass/cyst ____ A.S/A.D/A.U
Other ____

4. Oral Cavity

Appears normal ___
Tartar ____ Tooth # ____ Grade ____
Gingivitis ____
Mass/ulcer ____ Location ____
Periodontitis
Other ____

5. Musculoskeletal

Appears normal___
Lameness ___ (RF/RR/LF/LR)
Body condition score ____
Arthritis ____
Other _____

6. Abdomen

Appears normal ____
Tense/painful ____
Mass ____
Fluid ____
Organ enlargement ____
Other ____

7. Gastrointestinal

Appears normal ___
Vomiting ___
Diarrhea ___
Parasites ___
Other ___

8. Urogenital

Appears normal ___
Dysuria ___
Mammary masses ___
Prostate _____
Testicles _____
Prepuce/vulva ___
Pregnant ___Y/N
Other ___

9. Nervous System

Appears normal ___
Seizures ___
Ataxia ___
Other ___

10. Cardiovascular

Appears normal ___
Murmur _____
Pulse quality _____
RateRhythm _____

11. Lungs/Respiratory

Appears normal ___
Cough ___
Congestion ___
Dyspnea ___
Tracheitis ___
Other ___

Tentative Diagnosis
_____
_____
_____
_____
_____

Recommendations
_____
_____
_____
_____
_____

Sample Single Animal Record

Single Animal Record

Client Name: Farmer, Earl   **Address:** #32 State Road K, Anywhere, USA   **Phone:** 555-1111

Pet Name: Duke   **Species:** EQ   **Breed:** QH   **Sex:** M(C)   **Age:** 3 years   **DOB:** 2/22/2009

Color: PAL

MO	DAY	YR		Temp	MM	CRT
10	30	2011	Presented for wound to right shoulder, laceration from barbed wire, estimate 8–12 hours old. The wound is contaminated with grass, mud, and feces.	102°F	Pink	1 Sec
			Clipped and cleaned with chlorhexidine scrub, 5.45 mL xylazine sedation. Sutured 1.0 ethilon SI. 10.9 mL flunixin IV, 15 mL Penicillin IM. Stall confinement next 7 days, recheck 3 days. Rx-Flunixin paste 10g PO QD x 5 days. Pen injection 15 mL IM Q6H x 7 days.	103.0°F	Pink	1 Sec
11	1	11	Called owner to check on Duke. Owner says the horse is doing well and the wound looks like it is healing.			
11	3	11	Recheck- the owner has been unable to confine the horse as directed and he has been on open pasture. The wound is contaminated and the suture line had dehisced. There is necrotic tissue evident and purulent discharge. Debride tissue and flush with chlorhexidine solution. 5.45 mL xylazine IV, suture wound with 1.0 Ethilon SI. 15 mL penicillin IM. Strict stall confinement. Recheck 3 days.	105.0°F	Pale	2 Sec
11	4	11	Phone call to owner- Duke's condition seems stable, wound closure holding. Horse is eating/drinking, out in pasture. Told owner wound is still very fresh, horse is not ready for pasture activity. Confine to stall, give meds as scheduled.			
11	6	11	Arrived for recheck. Horse in turnout. Right front lameness apparent. Wound open and draining purulent material. Swelling of distal limb. Suspect cellulitis. Rec. Culture and sensitivity, wound care, debridement. O. declines, will monitor horse and continue with Banamine and penicillin. See AMA form.			

Sample Blood Results Form

Patient	Client	Date				
CBC	**Result**	**Ref. Range**	**Profile**	**Result**	**Ref. Range**	**U/A**
RBC		5.9–9.9	BUN		14–36	pH
Hgb		9.3–15.9	CREAT		0.6–2.4	Prot
HCT %		25–55	B.G.		64–170	Glu
MCV		37–61	K+		3.5–5.6	Ket
MCH		11–21	Na+		145–158	Bili
MCHC		30–38	Cl–		104–128	Blood
WBC		3.5–16.0	P–		2.4–6.2	Uro
EOS		0–1000	Alk Phos		6–102	S.G.
SEG		2500–8500	AST		10–100	Sediment
Bands			GGT		1–10	Cryst.
Lymph		0–800	Bili		0.1–0.4	RBC
Mono		0–350	Ca+		8.2–10.8	WBC
Baso		0–150	Amylase		100–1200	Casts
Platelets		100000+	Lipase		0–205	Bacteria
RBC morph			T. Prot.		5.2–8.8	Other
WBC morph			Albumin		2.5–3.9	
NRBC						**Fecal**
						Heartworm

Sample Anesthesia Monitoring Form

Anesthesia Monitoring Log

Pet Name: _____   Owner Last Name: _____

Date: _____   Surgery Performed: _____

Veterinarian: _____   Technician: _____

Laboratory

Body Wt. _____   Temp. _____   Pulse _____   Respiration _____

PCV _____%   TP _____ g/dL   BUN _____ mg/dL

BG _____ mg/dL   Other: _____

Medications

Drug Name	Strength	Amount (ml)	Route	Time	Initials

Fluid Maintenance

IV Catheter size _____   Location _____

Fluid type _____   Rate _____

Start fluid therapy _____   End fluid therapy _____

Total Volume infused _____

Induction

☐ Mask   ☐ Tank   ☐ Trach   ET tube size _____

Maintenance

☐ Mask   ☐ Tank   ☐ Trach

+ Pulse   O Respiration   ☐ **ET CO$_2$**   △ **SP O$_2$**   Blood Pressure: V Systolic   ^ Diastolic   X Mean

S Start   D End   E Extubated   A Anesthetic gas   O Oxygen

Sample Anesthesia Monitoring Form continued . . .

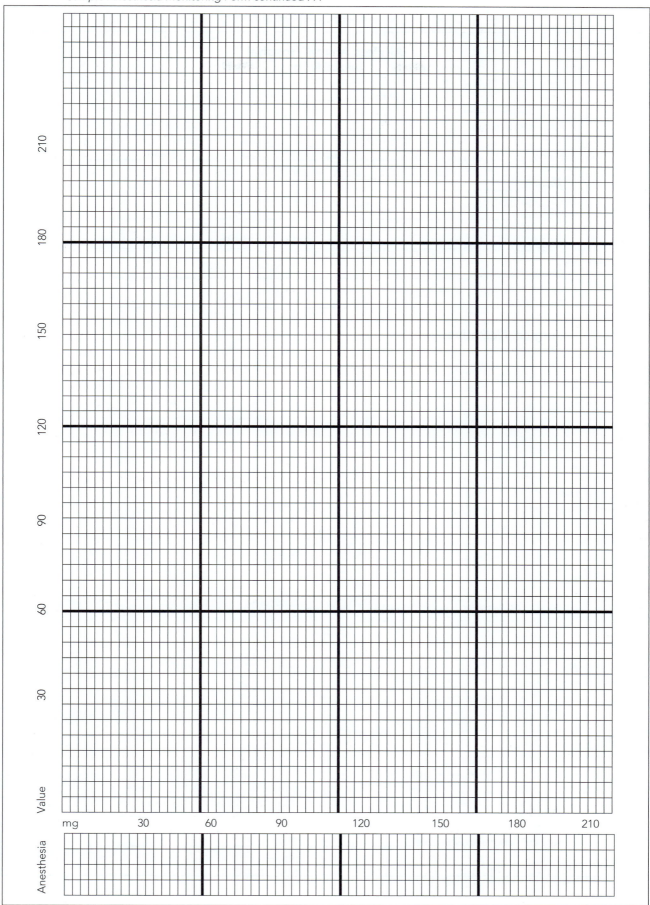

Sample Surgical Report Form

All Pets Animal Hospital
Surgical Report

Date _____

Patient Information:

Name _____ Age _____ Sex _____ Breed _____

Weight _____

Procedure _____

Surgeon _____

Surgical Technician _____

Anesthetist _____

Surgical Narrative:

DVM Signature _____ Date_____

Surgical Technician Signature _____ Date _____

departures from the standard protocol. These reports also include information about the types of instruments, suture, and suture patterns that the veterinarian used, as well as any notable events that took place during surgery.

7. Diagnostic imaging reports. Diagnostic imaging reports include information regarding the results of any radiographic or imaging procedures, as well as the actual radiographic images. The veterinarian, or in some cases a veterinary radiologist, will conduct an assessment of imaging studies. Common imaging reports include MRI, CAT scan, ultrasound, conventional and digital X-rays, and dental radiographs (see Sample Radiograph Release Form on page 192).

Clients may need to transport imaging studies to other hospitals for a second opinion. A radiograph release form is necessary to document that they have custody of the records.

8. Pathology reports. A veterinary pathologist issues a "path" report after the gross and microscopic analysis of tissues and other samples. Pathology labs analyze samples such as tumors, biopsies, and blood smears. The lab sends a written copy of the report to the clinic or hospital via mail, e-mail, or fax, and the pathologist is available for consult to discuss the diagnosis or prognosis (see Sample Pathology Report on page 193).

9. Referral and consult letters. Sometimes the veterinarian may refer a particularly challenging case to a specialist for a second opinion or additional diagnostics. The specialist will send a report to the referring hospital regarding the outcome of the visit. The specialist may recommend a treatment protocol or further diagnostics. For certain cases, especially those involving zoonotic disease, a variety of health care professionals, including physicians, may discuss the case if the pet's condition has health consequences for anyone in the owner's home.

10. Informed consent. "Informed consent" or "treatment authorizations" are necessary when a veterinarian offers to provide diagnostics, hospitalization, or treatment. When a client signs an informed consent authorization, it means that the veterinary staff has explained recommended services, what the expected outcome might be, and any potential risk. A veterinarian should not embark on a procedure, however minor, without the owner's consent. A signed form helps resolve disputes should any treatment or procedure become an issue or misunderstanding between client and veterinary hospital. It also prevents accusations by the client that the veterinarian performed services without his or her knowledge. Do not diminish the importance of keeping the client fully in the loop. Procedures you view as routine, such as shaving a surgical site or clipping matted hair, may cause distress in certain clients (see Sample Treatment/Consent Form on page 194 and Sample Surgical Consent Form on page 195).

11. Medical progress or SOAP (subjective, objective, assessment, and plan) notes. Medical progress notes are an organized method of arranging information to assure that the veterinary staff identifies and addresses all problems during a clinic visit or hospital stay and a treatment plan is implemented. You will more fully explore the concept of SOAP notes later in this chapter.

12. Discharge instructions. The clinic or hospital staff gives the client discharge instructions after a routine visit or just before releasing the patient from the hospital. Discharge instructions include a description of the performed services and procedures; complications that may arise; and explanations regarding feeding, exercise, and wound care, if necessary. The client also needs to know who to contact in case of emergency. Be sure to tell the client who will respond in case of an emergency. If an answering service or emergency facility handles after-hours calls for your practice, the client needs to know this before you discharge a pet. Review the discharge instructions in a quiet room before you bring the pet to the owner. Have the owner initial each individual instruction rather than providing a signature at the bottom of the page. This is documentation indicating that the client acknowledges understanding what he or she needs to do to care for the pet upon release (see Sample Discharge Instructions on pages 196–197).

Sample Radiograph Release Form

All Pets Animal Hospital

100 Any Street

Anytown, USA

Radiograph Release Form

Client's Name _____ Address _____

Phone _____ E-mail _____

I accept custody of _____ (number) original radiographs taken of my pet so they may be transported to Dr. _____ at _____ (name of veterinary hospital) for the purpose of consultation and/or additional veterinary care.

I understand that these radiographs are the property of **All Pets Animal Hospital** and that they will be returned directly to this hospital by the referral hospital or by me after they have been reviewed. I hereby accept the responsibility of returning these films to the facility. I accept that if I—or the subsequent veterinary hospital—fail to return these films, I am releasing the attending veterinarian and this veterinary practice of all legal liability for any charges of negligence, incompetence, or fraud that might be defended had these films been returned.

Pet's Name _____ Patient ID # _____

Doctor/facility expected to receive films _____

Number of films/CDs _____

Owner Signature _____ Date _____

Sample Pathology Report

Veterinary Pathology Lab, Inc.

222 W. 22nd Street, Hometown USA 10000

All Pets Animal Hospital   Tel: 555-555-1111
100 Any Street   Fax: 555-555-2222
Anytown, USA   Patient ID: 54321

Reference #	Doctor	Client	Pet Name	Received
APAH00100001	Newton	Plummer	Petie	11/12/2011

Species:	Breed:	Sex:	Age:	Reported:
Canine	Beagle	M(N)	1 yr	11/18/2011

Test: Histopathology
Source: 4 cm skin biopsy
History: Lesion of 30 days duration on the left thorax
Microscopic diagnosis: HISTIOCYTOMA
Prognosis: Favorable

Comments:
Cutaneous histiocytomas are reported to be the most common canine skin tumor, accounting for 20% of all skin tumors. The origin of the neoplastic cells is thought to result from abnormal proliferation of antigen-producing cells and may represent a unique hyperplastic, non-neoplastic process. Histiocytomas can occur anywhere on the body, but preferential sites include the muzzle, pinna, extremities, and scrotum. These tumors are most common in young dogs but do occur in dogs of all ages. Spontaneous regression is common. Tumors can be multiple, either concurrently or sequentially. Surgical excision should be curative.

This lesion appears to be completely excised.

Pathologist:
C. C. Ryder, DVM, PhD
Diplomate, American College of Veterinary Pathologists

Sample Treatment /Consent Form

Treatment Consent Form

All Pets Animal Hospital

Patient's Name _____ Description _____ Date _____

Owner's Name _____ Address _____

Contact Phone Number _____

I _____, owner of the pet identified above, have the authority to execute this consent.

I hereby consent and authorize the performance of the following procedures:

_____

_____

I understand that during the performance of the procedure(s), unforeseen or unfavorable outcomes may result. I release All Pets Animal Hospital for any and all liability.

I authorize the use of appropriate anesthetics and medications and the delegation of tasks to veterinary support personnel. I have been advised as to the nature and risks of the procedures. I understand results cannot be guaranteed.

I have read and understand this authorization and consent.

Signature: _____ Date: _____

Witness: _____ Date: _____

Sample Surgical Consent Form

Surgery/Anesthesia Consent Form
All Pets Animal Hospital

Owner/Guardian Name _____   Date: _____

Pet's Name _____ Species _____ Breed _____ Age _____ Sex _____
Procedure _____

I, the undersigned owner or agent of the owner of the pet identified above, certify that I am/am not (circle one) eighteen years of age or over and authorize the veterinarian of All Pets Animal Hospital to perform the above procedure(s). I understand that any anesthesia/surgical procedure carries with it a risk of complications. These risks, though not common, may include but are not limited to: allergic reactions, cardiac arrest, and/or death. Postsurgical complications may occur up to 10 days postsurgery and may include: damage to incisions by licking or chewing, breakdown of sutures due to overactivity, or infection at the incision site. It is my responsibility to monitor the pet after surgery and to seek medical advice as needed. All costs incurred are the responsibility of the owner/guardian, and All Pets Animal Hospital holds no liability for these charges.

To reduce the risk of anesthesia complications, all pets undergoing a procedure will have preoperative blood panels and physical exams done. An intravenous catheter will be placed and intravenous fluids delivered while under anesthesia. During the procedure, we will monitor the heart rate, blood pressure, pulse, respirations, temperature, oxygen, and carbon dioxide levels. **My signature on this form indicates that any questions I have regarding the following issues have been answered to my satisfaction.**
The procedure being performed:

_____

I am to receive discharge instructions detailing
How long it will take for my pet to recover
The most common complications
What to do in the event of a concern or emergency

Patients will be given pre- and/or postoperative pain medications and may possibly go home on continued medications, dependent on the procedure performed.

I verify that my pet has not eaten since before midnight prior to the procedure.

_____   _____
Signature of Owner/Guardian   Date

_____   _____
Signature of Witness   Date

Sample Discharge Instructions

All Pets Animal Hospital

100 Any Street

Any Town, USA

555-555-1111

Patient Discharge Instructions

Patient Name: _____ Client Name: _____ Date: _____

****IMPORTANT****

If your pet requires care or emergency treatment once he/she is discharged, it is the responsibility of the pet owner to seek appropriate care for the patient as soon as possible. Do not wait if your pet is distressed or otherwise in need of care. Please call Dr. Newton at (555) 555-1111. If you do not receive a reply in 15 minutes, please call the veterinary emergency clinic of your choice. **Initials of owner _____**

Behavior/Attitude:

_____ Monitor your pet for lethargy/depression, inappentence, vomiting or diarrhea, changes in appetite, and attitude

_____ Other _____

Diet:

_____ Feed your pet his/her normal diet.

_____ Other feeding instructions _____

Medications:

_____ Your pet was prescribed pain medication. Please give as directed. It is recommended that pain medication be given with food to avoid stomach upset. If your pet experiences any vomiting or diarrhea, or has blood in the vomitus/stool, discontinue the medication and contact Dr. Newton.

_____ Your pet was prescribed antibiotics. Please give as directed. It is important that a full course of antibiotic be given to prevent bacterial resistance. If your pet experiences an upset stomach, please call Dr. Newton.

_____ Other medications _____

Exercise:

_____ Keep calm/quiet with no strenuous exercise for the next days.

_____ Other _____

Sample Discharge Instructions continued . . .

Suture Removal:

_____ Check incision/wound daily for pain, heat, swelling, or discharge.

_____ Prevent licking or chewing at the incision/wound. An Elizabethan collar is recommended if the patient is bothering the wound.

_____ Sutures are absorbable and do not need to be removed.

_____ Other _____

Bandage:

_____ Your pet has a small wrap on its leg where an IV catheter was placed. It can be removed in 1–2 hours.

_____ Keep the bandage clean and dry. Wrap bandage in plastic wrap for protection in inclement weather. If the bandage becomes wet or soiled, a bandage change is recommended as soon as possible.

_____ Bandage should be reevaluated in _____ days

_____ Other _____

Fluids:

_____ Your pet has been receiving intravenous fluids. You may see increased urination for the next 24–48 hours

_____ Your pet has received subcutaneous fluids (fluids under the skin). A "hump" may be present that will disappear in time as fluids are absorbed. You can expect increased urination for the next 24–48 hours. The fluid "hump" may shift to the legs or chest.

Miscellaneous Instructions:

Client Signature _____ **DVM Signature** _____ **Date** _____

13. Necropsy notes. In the event of the patient's death, either in or out of the hospital, a client may request a necropsy to determine the cause of death (see the Sample Necropsy Form on pages 199–200). (You have heard this term referred to as *autopsy* as it applies to humans.) A complete external and internal examination takes place to look for evidence of the cause of death. You must record all findings and observations in the medical record. If the cause of death is obvious, the necropsy is complete. If the diagnosis requires further investigation, the veterinarian may send samples to a veterinary pathologist or toxicologist for review. You must also include this notation in the final report. When the veterinarian receives the report, he or she must call the owners and inform them of the findings. The veterinarian must note and report the conversation in the file record.

14. Authorizations for euthanasia and care of remains. Despite the emotional trauma and sensitive nature of the situation, if a client makes the decision to euthanize a pet, you *must* obtain a signed consent form giving the veterinarian permission to perform humane euthanasia *prior to the event* (see Sample Authorization for Euthanasia Form on page 201). Never direct your veterinarian into a room to begin euthanasia without signed consent, unless your clinic has a policy stating otherwise. Most clients know that they are required to give signed consent. You must make it as easy as possible by your kindness and concern and in the tone and delivery of your message. "Mrs. Jones, please forgive me, but I must ask you to complete this authorization form. If you can sign at the *X*, I'll send in Dr. Newton right away."

HOSPITAL LOGS THAT SUPPORT THE MEDICAL RECORD

Included in the array of records are also hospital forms and logs. Hospital logs are an extension of the medical record, but clinics typically keep them in a location close to where they are most commonly used. For example, **controlled drug logs** are usually kept in the same vicinity where the controlled substances are stored (see the Sample Controlled Drug Log on page 202). This helps maintain compliance by serving as a reminder to make the entry at the time you dispense the drugs. Federal law requires practices to keep a log of the use of *every* controlled drug that they dispense, dose by dose. You will also make a notation in the patient's record.

Another log that you usually find in its own location is the **radiology log** (see the Sample Radiology Log on page 203). You will have many patients that require radiographs. To produce the best-quality images, you must take careful patient measurements so that you can set the machine for proper exposure. The radiology log is the record of your patient's measurements and the X-ray exposure technique that you used to produce the image. This log is very useful for patients who need multiple images. You can refer to the log and program the machine at the right settings prior to the procedure, which saves time and stress on the patient.

Your hospital may also keep **maintenance logs** for equipment and the facilities (see Sample Hospital Maintenance Log on pages 204–206). Veterinary equipment is expensive and is subject to considerable wear and tear. Equipment such as the anesthetic machine, radiograph processor, autoclave, ultrasonic instrument cleaner, and ultrasonic dental scalers/polishers are examples of equipment that need regular maintenance. Logs let the hospital team know when equipment has been serviced and when safety checks were performed. Maintenance logs may also involve the functional parts of the building and property. Cleaning the restrooms, changing the furnace filters, and sweeping the walkways are examples of regularly scheduled chores. While we do not document maintenance chores in the medical

Sample Necropsy Form

All Pets Animal Hospital
Gross Necropsy Report

Date _____

Client _____ **Patient** _____ **Age** _____ **Sex** _____ **Breed** _____ **Weight** _____

Presentation: Died in hospital _____ **Died at home** _____ **Other location** _____ **Found by owner** _____

History:

_____

_____

Gross Findings:

External Appearance:

_____

_____

Head

Eyes _____

Ears _____

Nose _____

Mouth _____

Oropharnyx _____

Teeth _____

Other _____

Peripheral Lymph nodes

Submandibulars _____

Prescapular _____

Inguinal _____

Popliteal _____

Other _____

Thorax

Pleura _____

Diaphragm _____

Pericardium _____

Heart _____

Lungs _____

Trachea _____

Esophagus _____

Blood vessels _____

Other _____

Sample Necropsy Form continued . . .

Abdomen

Diaphragm _____

Liver _____

G. bladder _____

Spleen _____

Pancreas _____

Kidneys _____

Ureters _____

U. bladder _____

Stomach _____

Intestine _____

Omentum _____

Peritoneum _____

Lymph nodes _____

Other _____

Integument

Hair coat _____

Skin _____

Nail beds _____

Mucocutaneous junctions _____

Special Testing:

Tissues submitted for:

Histopathology _____

Culture _____

Toxicology _____

Rabies _____

Other _____

Comments:

_____

_____

_____

_____

_____

DVM Signature _____ **Date** _____

Care of remains

_____

_____

Sample Authorization for Euthanasia Form

All Pets Animal Hospital

100 Any Street

Any Town, USA

AUTHORIZATION TO PERFORM EUTHANASIA

Client Name _____ *Patient Name* _____ *Species* _____ *Breed* _____

I, the undersigned, do hereby certify that I am the owner or duly authorized agent for the owner of the animal described above, that I do hereby give All Pets Animal Hospital permission for humane euthanasia and release its staff from any and all liability.

I acknowledge that Dr. Newton has met with me personally and discussed the euthanasia of my animal. I also certify that to the best of my knowledge, said animal has not bitten any person or animal during the last fifteen (15) days, and has not been exposed to rabies. I further understand that I assume financial responsibility for all services rendered.

PLEASE INDICATE YOUR DECISION FOR CARE OF REMAINS BY INITIALING BELOW:

_____ Owner takes remains

_____ Communal cremation

_____ Private cremation (remains will be returned to All Pets Animal Hospital for pickup by the owner)

_____ Please hold remains pending our decision. (10 day limit, after which All Pets Animal Hospital will arrange for communal cremation)

Learning the cause of death of your pet can be of great help in relieving the suffering of other pets, as well as contributing to our understanding of health and disease.

_____ I DO/DO NOT authorize a postmortem evaluation.

Date _____, 201_____.

_____   _____

Signature   Witness

Sample Controlled Drug Log

Veterinary Controlled Drug Disposition Record

Name of Drug _____ Form _____ Strength_____ Size _____

Date	Time	Last Name, Patient Name	Breed	Signature of Person Dispensing	Amount Used	Balance on Hand

Sample Radiology Log

Radiology Log

Date	Pet Name	Owner Last Name	View Taken	Measurements (cm)	Lateral kVp, mA, sec	V/D kVp, mA, sec

Sample Hospital Maintenance Log

All Pets Animal Hospital
HOSPITAL MAINTENANCE LOG
Date_____

Hospital Area	Chore	Date	Initials
Clinic Entrance	Clean entry mat		
	Sweep walkways		
	Clean doors/windows		
	Pick up trash		
Reception Room	Vacuum/mop floors		
	Dust		
	Straighten counters		
	Clean windows		
	Straighten magazines		
	Straighten retail merchandise and restock		
Exam Rooms	Restock drawers		
	Re-charge oto/ophthalmoscopes		
	Vacuum/mop floors		
	Spot clean walls		
	Clean sinks		
	Empty trash		
	Replace air freshener cartridge once a month		
Washroom	Vacuum/mop floor		
	Clean sink		
	Clean toilet bowl		
	Wipe down fixtures/mirror		
	Restock paper towels and toilet paper		

Sample Hospital Maintenance Log continued . . .

Parking Grounds	Reception Room	Examination Rooms	Employee Lounge
Clear debris	Vacuum/mop floors	Vacuum and mop floors and under tables	Turn off appliances (nightly)
Sweep walks	Dust	Vacuum chairs	Wash dishes
Empty trash	Straighten displays Disinfect all counters	Spot clean walls & floor	Vacuum/clean floor
Clean mats	Wipe down table and organize books	Clean sinks	Clean microwave
Wash windows	Clean baseboards	Empty trash containers	Clean & dust table(s) & chairs
Snow removal	Clean walls	Restock and organize drawers –	Throw away old food from refrigerator
	Restock treats	Restock paper towels, toilet paper	Washroom
	Clean windows	Clean otoscope cones	Check toilet paper and restock
	Wipe down windowsills	Handouts, puppy/kitten kits	Clean counter
	Water plants	Clean scale mat	Vacuum and mop
	Clean retail shelves	Replace air freshener cartridge	Clean toilet bowl inside and out
			Wipe down mirror and light fixtures

Sample Hospital Maintenance Log continued . . .

Radiology Area	Surgery Area	Kennel Room	Prep/Treatment Area	Storage Area
Vacuum/scrub floors	Disinfect counters/ surgery table	Clean /scrub kennels	Vacuum & clean floors	Laundry
Empty & clean trash containers	Vacuum/scrub floors	Wash kennel dishes and litter trays	Clean utensils and instruments ASAP	Sweep/mop floors
File X-rays	Spot clean walls	Keep food tubs filled	Clean prep sinks	Disinfect/main- tain isolation kennel
Ensure red light off	Check dates/ restock emer- gency drugs	Vacuum & disinfect floors, under and behind kennels	Empty & clean trash containers	Clean/organize cabinets & tables
Tidy and orga- nize counter	Clean/restock pack storage areas	Organize & restock food and supplies	Check fridge/cup- boards for outdated inventory	Organize pre- scription pet food
Clean X-ray table	Replace soda- sorb granules	Wash & dry all laundry	Refill soap dispenser(s) and stock solutions	Inventory back stock
Check/order supplies	Inspect all surgical instru- ments used	Empty & clean trash containers	Wipe down cup- board doors and counter tops thoroughly	Clean freezer
Clean & straighten apron & gloves dust & straighten files	Clean surgery table	Clean tub & surrounding area	Spot clean walls	
Replace devel- oping solution	Replace cold sterilization solutions	Scrub cage doors	Check inventory/ order	
Replenish developing solution	Inventory supplies	Clean windows	Clean refrigerator	
Clean baseboards			Clean autoclave	

record, they do provide evidence of a patient's care. There are situations that can arise where it may become necessary for the clinic to prove that it properly maintained equipment.

Because of the large number of required forms and logs for documentation of the patient record and the hospital's activities, many clinics purchase standard, preprinted forms. Veterinary websites often contain "forms libraries" where practices share a compilation of forms and logs. Because they are in the form of Word documents, clinics can customize them to fit their clinic's needs.

You are beginning to understand the extent of the information that you need to document in the medical record and in documents that support the medical record. You will need to enter notes and file information into records every day. Keep this task as current as possible. Records that stack into piles waiting for doctor's notes, surgical reports, or other entries are less likely to be accurate and may not be up to date when another member of the hospital needs the record.

HOSPITAL OR CLIENT: WHO OWNS THE MEDICAL RECORD?

All documentation that comprises the medical record is the property of the veterinary hospital. Each veterinary facility must retain such records for a time period that the state veterinary medical board or other licensing agency determines. At any point within the mandated time frame, the hospital must be able to produce the medical record for review by the state, the owner, the veterinary medical board, or another qualified requester. Sometimes medical records are needed for unexpected reasons. There may be a dispute between neighbors, there may be debate about animal ownership, or the owner's physician may want to review the record if there is a suspected zoonotic illness. The owner may receive any part of the medical record or request that the clinic send specific information within the record to a third party. You should obtain a signed consent form before you release any part of the record. In the case of co-owned animals, it is important to obtain consent from both owners before you share any information.

The Medical Record Is Confidential

Veterinary patient medical records are confidential. No hospital employee should divulge any of the information within the record to anyone outside of the veterinary team. This can happen easily and often just because you are well meaning. Let's look at an example.

Mrs. Brady is leaving the hospital in tears. She's just been told that her precious dog Bridget has a tumor requiring removal. There is a high suspicion that it is cancerous and may have spread. As Mrs. Brady is checking out, Ms. Stevens is checking in. She is speaking to Holly, the veterinary technician. Ms. Stevens says, "Oh, I'm so sorry. Is that lady okay?" Holly leans over the desk and whispers, "It's really sad. Looks like her dog may have cancer."

We know that Holly is just trying to be sympathetic and satisfy normal human curiosity, but she has violated Mrs. Brady's confidentiality. Ms. Stevens and Mrs. Brady do not know one another and may never cross paths again, but Holly has clearly overstepped a boundary. *Never discuss any part of a patient's medical history with anyone but the owner or his or her authorized representatives.* Even offhand remarks can cause consequences down the road.

If you find yourself in a position where a client asks you to remark on a patient's condition, simply reply (in our example), "I'm sorry, Ms. Stevens. I'm not allowed to discuss another patient's condition. We practice that for all of our clients." You've now accomplished two things: you've protected the confidentiality of the patient and client, and you have earned Ms. Stevens's trust. She knows that she will never have to worry about disclosure of her private matters. Conditions that are part of our everyday life in the veterinary hospital can be a source of embarrassment for clients. People can be sensitive when their pets are diagnosed with common problems such as parasites. They often think that it reflects poorly on their ability to care for a pet. Even if the problem is as simple as a case of fleas, you cannot disclose it.

Consider this scenario: Holly is at the front desk. Mrs. Fisher calls. Her cat Sammy has been scratching. Holly asks her if she has seen any fleas. Holly remarks that Mr. Connor, who lives close to her, had to treat his dog, house, and yard for a flea problem. Mrs. Fisher, who uses monthly flea preventative and is a very meticulous housekeeper, makes a mental note to keep her children from going over to the Connor house to play. She does not want them bringing fleas home or being exposed to the pesticides that Mr. Connor is applying to his lawn. Remember, your clients can tell each other anything that they want about their pets or their care, but the hospital must *never* violate their privacy.

The Medical Record Is Portable

From the time of entry until the time of discharge, the medical record follows the patient through the hospital. The record may start out in a wall pocket outside the exam room door or on a clipboard on the front desk. When the patient changes locations in the hospital, the medical record goes with it. This insures that all hospital personnel can write their notes and comments in a timely fashion. This decreases errors and improves communication between staff members so that everyone knows what has transpired. For example, the veterinarian orders medication for an eight-week-old puppy that he has just diagnosed with roundworms. The owner authorizes treatment and leaves the puppy at the clinic so that he can run a quick errand. The doctor writes the order in the record and hands the puppy and the medical record to Holly. Holly gets the medication and wraps it in a small treat; the puppy eats it willingly. Holly then kennels the puppy until its owner comes back to pick it up. The owner is later than expected, and Holly has gone to lunch. The office manager checks the record and sees that Holly has checked and initialed the pharmacy order for the puppy and made a notation in the record that she observed the puppy consuming the medication. Now there is no question that the clinic has treated the puppy appropriately, and the receptionist at the front desk can confidently discharge the pet to the owner. She invoices the charges from the notations in the records and files the account information in the appropriate place. The receptionist then puts the file in its designated place, where she will easily locate it when the puppy returns in two weeks for his next visit.

GOING GREEN: THE PAPERLESS PRACTICE

Medical records are going green. Many veterinary practices are transitioning to paperless operations. Keeping digital information has many advantages. It cuts down on the expense of office supplies, conserves resources, and eliminates the need for space to file and maintain records. It also eliminates one of the more labor-intensive tasks of the front staff, as filing and maintaining records is often a tedious job. The veterinary software that supports keeping medical records will also help keep information organized and easier to locate. In most cases, the software cross-references information so you can locate a record even if you only have the pet's name. It also completely eliminates the problem of illegible entries (see the Sample of Illegible Handwriting on page 209). If a staff member has unreadable handwriting or if

Sample of Illegible Handwriting

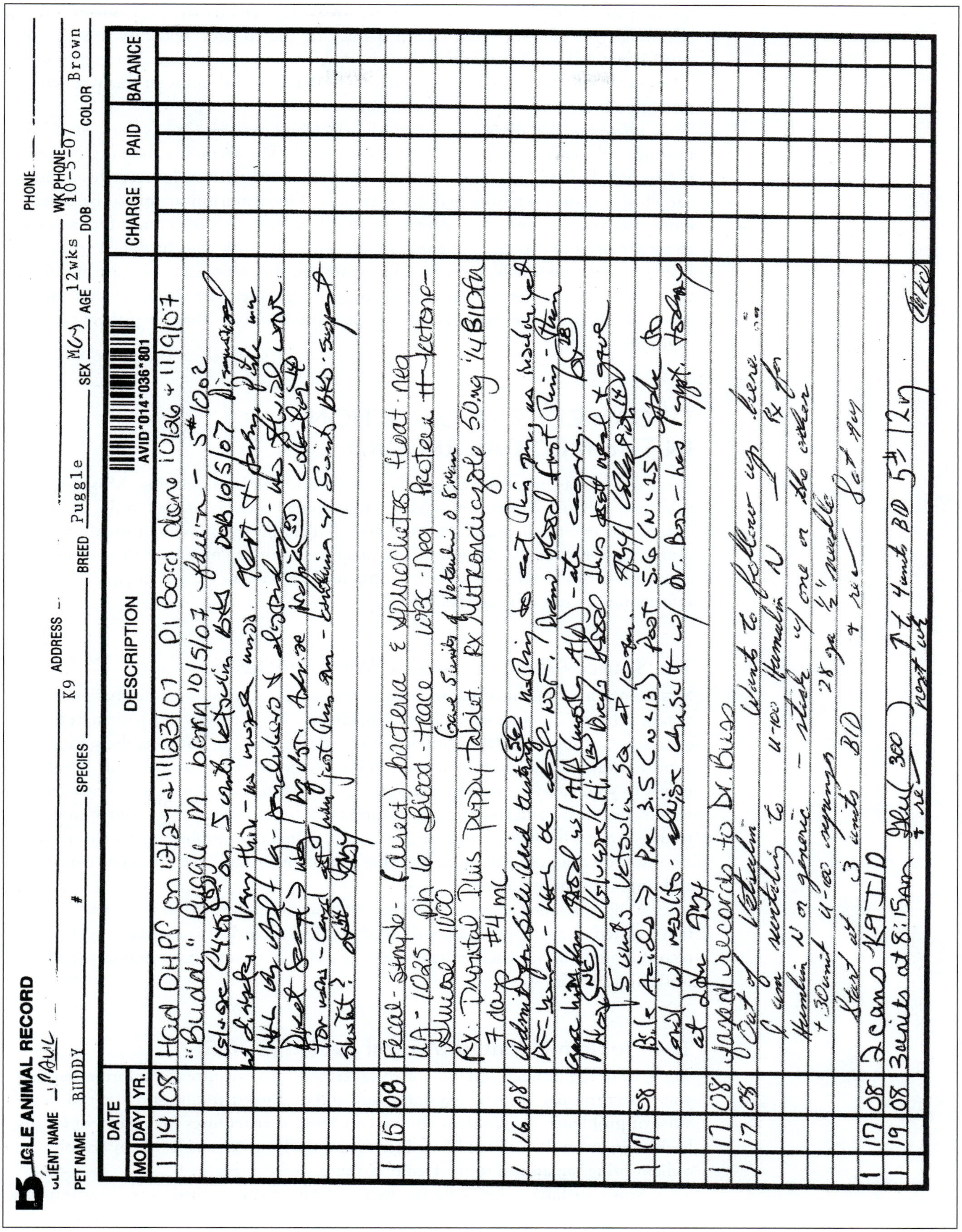
BEAGLE ANIMAL RECORD
CLIENT NAME
PET NAME BUDDY
SPECIES K9
ADDRESS
BREED Puggle
SEX
AGE 12wks
DOB 10-5-07
COLOR Brown
PHONE
WK PHONE
AVID*014*036*801
DATE
MO DAY YR.
DESCRIPTION
CHARGE
PAID
BALANCE

spelling skills are poor, electronic records will help remedy the problem. Another advantage of a computerized record is tracking the charges. It is easy to overlook small items and services that the clinic has used in the patient care. Veterinary software programs allow you to record charges so that the final bill is an accurate reflection of your services.

There are also disadvantages to the electronic record. One is the cost of setting up and maintaining the veterinary software. The second is the portability of the record. Unless there are sufficient computer workstations located throughout the clinic, team members may have to access information at locations distant from the patient. Lastly, like anyone who has worked with computers knows, there are only two types of people in the world—those who have lost information and those who will! It is crucial to maintain the veterinary software and back up and store the information so it is always accessible. Any situation that causes a loss of electrical power may also compromise hospitals that keep their records digitally.

ORGANIZING INFORMATION IN THE MEDICAL RECORD

Every doctor has his or her own style and method of record keeping. Most learn in veterinary school a format for making and keeping organized notes. As doctors leave school and enter real-world practice, their dedication to following formalities may diminish. Some may be organized and precise, and others may have their own ideas about what constitutes sufficient information. In general, we can organize medical information in one of two ways.

The Source-Oriented Medical Record (SOMR)

Source-oriented medical record refers to the process of recording information as you obtain it from "the source." In human medicine, this is usually the patient, a friend or relative, or collected data from diagnostic testing. For our patients, the source will almost always be the animal's owner. You write down the information as the client tells it to you, and you develop a chronological record. You also include notes on lab results, diagnosis, and treatment as information becomes available. Source-oriented records are less formalized and are often less detailed than problem oriented medical records. You can keep data from multiple pets on one record, which can be confusing and time consuming to sort through.

A more detailed approach to organizing medical information is the use of a system referred to as the **problem-oriented medical record**. It is the standard form of record keeping in human medicine and is usually the format that veterinary teaching institutions prefer.

The Problem-Oriented Medical Record (POMR)

The goal of the problem-oriented medical record is to organize the information that you collect through history, physical exam, and diagnostic testing and use it to create a list of problems. A "problem" is any condition that requires management by the health care team. It is a broad-based term, and the hospital uses it for any reason why a client brings an animal to the facility for care.

You will have a key role in the creation of this record, as it is often the veterinary technician who begins the process of taking a history prior to the veterinarian entering the exam room. You should become comfortable with the POMR format.

Signalment and History Earlier in the chapter, we introduced the concepts of signalment and history. Do not forget the critical importance of this information. Think of it as an introduction. When you are out socially, the first thing you say to someone you are meeting for the first time is, "Hi, my name is _____." Our name is the first information that we give to people. As the conversation progresses, the two of you reveal more facts about yourselves. You may discuss where you grew up, what you do for a living, and your interests. The signalment is our introduction to the patient.

Before reading further, practice forming your own signalment statement from the client record below. Refer back to pages 179 and 182 for review.

Single Animal Record

Client Name: White, Rochelle **Address:** 100 State Street, Anywhere, USA
Phone: 555-1111 **E-mail:** Rwhite@email.com

Pet Name: Roxy   **Species:** K9   **Breed:** Dachshund   **Sex:** F(S)
Age: 12 years   **DOB:** 10/11/99   **Color:** Black/Tan

MO	DAY	YR		Temp	MM	CRT
11	15	2011	Presented for weight loss, 3 weeks' duration. Wt. 9.5 lb.	102.4°F	Pink	1 Sec

---

Your signalment statement should be "Roxy is a twelve-year-old female spayed Dachshund."

Let's restate why this information is so important. It is essential because it provides a framework for the veterinarian to evaluate the patient before he or she even sees her. Using the example above, Roxy is in the clinic because she is experiencing weight loss. You've weighed the dog and taken vitals. You let the owner know that the veterinarian is ready to see the patient. Knowing that the patient is a canine helps the doctor start to formulate a list of possible causes for the weight loss. Maybe the pet has parasites or a poor diet. It could be kidney disease or even a bad tooth that makes it painful for Roxy to eat. If you include the signalment in the patient's description, the veterinary team is able to more accurately predict the nature of the weight loss. This particular pet is a twelve-year-old female (spayed) Dachshund. This information really changes the way we think about this pet's weight loss, because we now have to consider causes in animals that are geriatric.

Before reading further, let's refine our statement about Roxy's visit to the hospital. We can now state, "Roxy is a twelve-year-old female spayed Dachshund presented for a history of weight loss of three weeks' duration."

As Roxy's signalment and history evolve, the next step in creating the POMR is to document the physical exam findings. You will list and address separately each of Roxy's problems. You then will combine the physical exam findings with data from the laboratory to round out the problem list.

Single Animal Record
Client Name: White, Rochelle **Address:** 100 State Street, Anywhere, USA **Phone:** 555-1111 **E-mail:** Rwhite@email.com

Pet Name: Roxy   **Species:** K9   **Breed:** Dachshund   **Sex:** F(S)
Age: 12 years   **DOB:** 10/11/99   **Color:** Black/Tan

MO	DAY	YR		Temp	MM	CRT
11	15	2011	Presented for weight loss, ravenous appetite. Drinking more water/urinating frequently. Wt. 9.5 lb. Patient is lethargic, Eyes- cataracts O.U. Ears- OK, Teeth- grade 2 dental tartar, Lymph nodes- OK, Hydration- OK, Heart/Lungs- OK murmur, Abdomen- soft, nonpainful, Musculoskeletal- weight loss of 2 lbs., Integument- OK, Urogenital- OK	102.4°F	Pink	1 Sec

Now, let's take the data from the physical exam and laboratory and make a list of Roxy's problems. We are going to call this the **Master Problem List**, and it is usually located at the front of the record.

Master Problem List

Example Master Problem List
Roxy K9 White

1. Weight loss
2. Ravenous appetite
3. Drinking/urinating frequently
4. Grade 2 dental tartar
5. Cataracts O.U.

Once you have established the Master Problem List, the veterinarian will rank Roxy's problems according to their severity. For example, in light of Roxy's weight-loss problems, her cataracts are probably not going to be at the top of the list for evaluation and treatment. That can come after the veterinarian solves the weight-loss and appetite problems.

SOAP The second step in formulating a POMR is to SOAP each problem. In this case, **SOAP** does not refer to bathing your patient! It is an acronym for **subjective, objective, assessment, and plan.**

Subjective The **subjective** component of the POMR consists of information that most often comes from the owner's observations and description of the problem. The owner's perception of what is happening is often quite different than what is actually happening to the pet. The layperson can easily misinterpret the clinical signs the pet exhibits or miss subtle findings. It is important to *listen* carefully and write down the information exactly as the owner describes it. It is the veterinary team's job to sort through this information and identify the problems.

Sometimes the owner is unable to provide much information. The owner simply knows that his or her animal is not right but may not be able to say why. In this case, you will help the owner recognize abnormalities with open-ended questions. We do not want to lead the client into an answer. For example, when speaking to Roxy's owner, we want to ask about the amount of water she is consuming. You may phrase the question so that the owner provides unbiased information.

INCORRECT: "I see that Roxy has lost two pounds. Has she been drinking more water?"

CORRECT: "I see that Roxy has lost two pounds. Tell me about her water consumption. Has it changed at all?

In the first statement, we are leading the client into an answer by making a suggestion before she has had a chance to think about it. The owner may think that she is aiding in the diagnosis by answering positively. In the second statement, the client must think about and evaluate Roxy's water consumption before answering. (Did you know that technically animals do not have *symptoms*? A symptom is a condition that the patient verbalizes. Our patients cannot talk about their illnesses, so we call their conditions "clinical signs.")

Objective Once you obtain the subjective information, the **objective** exam begins. Objective data refers to that which you can measure and substantiate. For example, Roxy's owner reports that her dog "feels hot." This is a subjective statement. The owner notes the change, but it really may not be accurate. You take a temperature and record that the pet is normal at 102°F. That is objective. It is not something that you "feel" is happening—you have proved it with a thermometer reading. Most of the objective data will come from the physical exam and laboratory findings. The veterinarian will use these data to formulate the assessment.

Assessment The patient **assessment** is a list of the possible causes based on the findings of the physical exam and lab data. For our friend Roxy, who has a ravenous appetite and is still losing weight, we will need to come up with a list of possibilities that cause that set of clinical signs. This is the veterinarian's responsibility, but you will come to recognize the clinical signs of many common diseases for many animal species. This type of experience makes you an invaluable member of the team, because you will be able to anticipate what the doctor wants in many situations.

In Roxy's case, the doctor will make a list of her clinical signs and then formulate the "rule-outs." The veterinarian will want to *rule out* the list of possible causes until he or she reaches only one cause, the *diagnosis*.

Roxy's clinical signs:

1. Ravenous appetite
2. Weight loss
3. Increased water consumption and urination (owner confirms)

Roxy's rule-outs (R/O):

1. Diabetes mellitus
2. Parasitism
3. Kidney disease
4. A combination of any of the above (or maybe something totally different)

All three of the conditions listed above are common causes of Roxy's clinical signs. How do we figure out which one is the cause? How do we get to the diagnosis?

Plan If we are going to figure out Roxy's problem, we need a plan. Sometimes we can identify the problem just by a physical exam, which helps us formulate a plan of how to treat it. In Roxy's case, we have to rule out three possibilities. We will need to do some testing.

Plan for Roxy:

1. Fecal exam
2. Complete blood count (CBC)
3. Urinalysis
4. Complete blood profile

The veterinarian has formulated the plan and made notes in the medical record that these are the diagnostic tests that he or she needs. As the technician, you will often be responsible for the collection and processing of these samples. Let's look at Roxy's results.

Roxy White	Lab Report Form	Comments
Exam		
Fecal	No ova seen, no parasites on float or direct exam	negative
Urinalysis	Voided sample, appearance- clear, pH-7.0, glucose- +2000, ketones- 1+, blood- neg, leucocytes- neg, protein- neg, Sediment- neg, specific gravity- 1.030	glucosuria, ketonuria
CBC	Within normal limits	
Blood profile	Blood glucose 450mg/dl	hyperglycemia
Thyroid (T4)	Within normal limits	

Based on the results of the subjective/objective data, the veterinarian makes a presumptive diagnosis of diabetes mellitus. The veterinarian will decide if he or she requires any further testing to support the diagnosis. There are additional blood tests available that can determine the best course of therapy for Roxy.

By convention, each separate problem on the Master Problem List is SOAPed. This serves to organize and create a detailed record that addresses every exam finding so that we do not ignore anything. Often, the veterinarian will "retire" problems with lesser impact on a patient's health, until a more appropriate time. A plan that the veterinarian formulates based on history, clinical, laboratory, and imaging data is a compact and organized approach to reaching a diagnosis.

SUMMARY

The medical record is the cornerstone of good patient care. Properly documented records provide accurate and current information about the client and patient, fulfill the legal requirement for maintaining medical records, and protect the hospital if litigation occurs. As the veterinary technician, you will be responsible for making entries in the record and documenting the care and treatment that you provide.

Take record keeping seriously. On busy days it may feel like a burden, and some notations may seem inconsequential. Recording anything that adds to the record's accuracy is good practice, and it is impossible to know when a small entry may have a big impact.

As a student, use the medical record to help you organize your thoughts. Practice forming signalments and histories with your personal pets and the animals you encounter through your classes. Use the SOAP format to help you categorize the information you obtain. You will begin to see patterns emerge that will help you recognize common conditions and illnesses.

Lastly, remember . . . if you didn't write it down, it didn't happen.

TEST YOUR KNOWLEDGE

1. Of the following, the *most* important reason for maintaining medical records is that
   a. it is a legal requirement.
   b. it provides a basis for patient care.
   c. it helps keep track of client accounts.
   d. it helps identify disease patterns.

2. All medical information should be written in pencil to make it easier to correct errors.
   a. True
   b. False

3. On average, how long are medical records stored?
   a. 3–6 months
   b. 1–3 years
   c. 3–5 years
   d. 5–7 years

4. What are the two primary ways to organize information in the medical record? How do they differ?

5. What does the acronym SOAP stand for? Describe the information represented by each letter.

6. What are the advantages and disadvantages of paper versus computer records?

7. Which of the following documents contains a complete description of X-rays, MRIs, CAT scans, ultrasounds, and other imaging?
   a. Surgical report
   b. Radiology report
   c. Anesthetic record
   d. Discharge instructions

8. During a spay procedure, the veterinarian finds a small mass on the skin that is ulcerated and should be removed. Before proceeding, what is the best course of action?

9. List five reasons why veterinarians keep medical records.

10. Mary Rutherford calls your practice asking for information about Jack Bradshaw, a two-year-old Labrador retriever. Mary Rutherford and Jane Bradshaw are best friends. She wants to know the results of Jack's hip X-rays because she is taking one of the puppies from a litter that he sired. What is your appropriate course of action?

BIBLIOGRAPHY

Veterinary Medical Data Base. 2010. "Home page." Last accessed December 9, 2011 from the Veterinary Medical Data Base website.

Glossary

American Association of Laboratory Animal Sciences (AALAS) "An association of professionals that advances responsible care and use of laboratory animals to benefit people and animals."

American Association of Veterinary State Boards (AAVSB) An association serving veterinary licensing boards by providing programs such as testing and the approval of continuing education programs.

Agribusiness Farming or the production of agricultural products by companies or corporations.

Anesthetic report A portion of the medical record that details all the events of an anesthetic procedure.

Animal guardianship A term used to refer to the elevated status of a pet as a member of the family rather than property.

Animal industry Business and activities that derive a profit from the use of animals or the sale of animal products, services, and supplies.

Animal ownership A term used to refer to the status of a pet as property of an owner but entitled to humane care and legally protected from abuse.

Animal welfare Providing for the physical, psychological, and social needs of an animal.

Animal and Plant Health Inspection Service (APHIS) An agency of the U.S. Department of Agriculture, its mission is to protect animal and plant resources from pests and diseases.

Assessment The portion of the problem-oriented medical record (POMR) that details the clinical findings of the history and physical exam, including any diagnostic rule-outs.

American Veterinary Medical Association (AVMA) A nonprofit association representing veterinarians. "The objective of the Association shall be to advance the science and art of veterinary medicine, including its relationship to public health, biological science, and agriculture."

Biohazard protocols A written protocol detailing the actions needed for containing biohazards. Common biohazards include the handling of needles and other sharps, animal waste management, and proper disposal of expired drugs.

Biomedical research Experiments conducted to further the body of scientific knowledge in both human and veterinary medicine. Biomedical research may incorporate the use of animals as models for disease or to test the effectiveness of drug therapy or treatments.

Biotechnology The use of a living organism or system in the study of science or medicine.

Body language Gestures, postures, and other nonverbal signals utilized for communication in both animals and people.

Breed Association A group of people sharing a common interest in a specific breed of animal. Breed associations often set standards of appearance and performance and may host events promoting responsible husbandry and ownership.

Care of remains The manner in which an animal's body is cared for after death. Private cremation, communal cremation, and burial are options.

Center for Veterinary Medicine A division of the Food and Drug Administration (FDA) that regulates medications and additives given to animals.

Cloning Producing genetically identical individuals from a single common ancestor.

Comfort zone A mental or physical boundary established by a human or animal that maintains emotional neutrality; a space that does not provoke anxiety.

Common law Decisions made by judges through trials rather than by statute.

Companion animal Any animal, regardless of species, that provides interaction as well as emotional, social, physical, and psychological support.

Confidential/confidentiality Refers to information that cannot be shared except with designated people; private information that may not be disclosed to parties not directly involved or affected.

Confined Animal Feeding Operation (CAFO) An area of confinement for the raising and feeding of animals for a period of forty-five days or more in a twelve-month period.

Conflict resolution The process of solving a problem between people using constructive and nonconfrontational techniques.

Consult room An area of the veterinary hospital where veterinary staff may have private discussions regarding a pet's condition, treatment plan, or quality of life.

Continuing education Courses, seminars, or other types of learning opportunities that offer up-to-date information for professionals once their formal education has been completed.

Contract farm A farm that has a mutual agreement with a buyer to provide an agricultural product at a specific quantity and price. Other conditions may also apply, such as the use of certain types of seed or fertilizers. The buyer may provide technical advice or other support.

Corporate farming (see *agribusiness***)** Farms or other agricultural operations that are owned and operated by a commercial business rather than a single family.

Committee on Veterinary Technician Education Activities (CVTEA) A program of accreditation for the training of veterinary technicians.

Drug Enforcement Agency (DEA) A division of the U.S. Department of Justice, the DEA enforces the laws that regulate the use of controlled substances.

Diagnosis The determination of the cause of a disease, confirmed through analysis of patient data such as history, physical exam, laboratory tests, and imaging.

Diagnostic imaging report   A report detailing the findings of any imaging, such as radiographs, ultrasound, magnetic resonance imaging (MRI), or computerized axial tomography (CAT scan).

Diagnostic Laboratory   A laboratory utilized to process biological samples to aid in the diagnosis of disease. Cytology, histopathology, bacteriology, and microbiology are examples of laboratory tests carried out in a diagnostic laboratory.

Dignity   The recognition and effort to meet a person's individual needs.

Discharge instructions   Written and verbal descriptions of how to properly care for a pet being released from the hospital.

Documentation   Any written record that conveys information regarding a client or pet.

Domestication   Adapted to human living conditions; when humans have chosen animals for breeding, to pass on desirable traits such as herding ability or milk production.

Ego   The recognition of yourself as a unique and separate human being with self-esteem and self-worth.

Embryo transfer   A technique of reproductive management; a process by which the fertilized embryos of a female are collected and transplanted into the uterus of another female.

Emergency hospital   A type of veterinary facility that sees only emergent cases or provides services when regular veterinary hospitals are closed.

Emotional bank   A reserve from which you derive the ability to meet the needs of others in a kind, caring, and respectful manner. Your emotional bank is replenished when your own needs are met by others and by what you do for yourself.

Empathy   The ability to understand another person's feelings, experience, or situation.

Environmental Protection Agency (EPA)   A federal agency created to protect the environment and human health by enforcing laws and regulations designed to safeguard air, land, and water quality.

Equine practice   A type of veterinary facility limited to the medical and surgical care of horses and related equids.

Ethically conscious   The principle that dictates that your behavior and professional practices will be held to a high standard whether you are being observed or not.

Ethics   The right or proper thing to do, as determined by an individual's religious, cultural, and social values.

Euthanasia   A humane method of a quick and painless death.

Exam room   A designated area of the veterinary hospital used to perform physical assessments and client consultations. The room is equipped with common instruments and supplies to facilitate a physical exam and minor procedures.

Exotic animal   Any companion animal that is not a dog, cat, horse, or cow. Common examples are hamsters, guinea pigs, birds, reptiles, arachnids, and amphibians.

Exotics practice   A type of veterinary practice that is either limited to or includes exotic animals.

Fair hiring practices   Laws that prohibit discrimination based on gender, race, religion, or national origin.

Family farm   Farms owned and operated by the members of a single family, often passed down from generation to generation.

Food and Drug Administration (FDA)   A government agency that protects the public health by assuring the safety of drugs and medical devices as well as the drugs and additives used in animals.

Feline practice   A type of veterinary practice limited to the medical and surgical treatment of cats.

Financing   The process of obtaining funds for purchasing a business, equipment, and supplies. Money may be obtained through banks, personal loans, grants, endowments, or other sources.

Food Safety and Inspection Service (FSIS)   The division of the U.S. Department of Agriculture (USDA) that is responsible for ensuring the safety of the nation's meat, poultry, and egg products.

Gene knockout technology   The inactivation of a gene in a sequence of DNA.

Genetically modified organism (GMO)   An organism that has been altered at the molecular level to introduce or eliminate genes in a sequence of DNA.

History   The portion of the problem-oriented medical record (POMR) that describes the reason for the patient's visit to the veterinary clinic and what clinical signs have been observed by the owner.

Hospital policies and procedures   A manual describing how various problems, situations, and circumstances are handled within the framework of the veterinary practice.

Hospital ward   An area of the veterinary hospital where patients are housed for treatments, recovery, or hospitalization.

Human-animal bond   Describes the relationship between animals and people that is emotionally and psychologically beneficial for both.

Informed consent   A description of offered services, the risks involved, and the expected outcomes of any treatment.

Interpersonal communication   A set of verbal and nonverbal behaviors that are utilized to convey thoughts and feelings between people.

Isolation   An area of the veterinary hospital in which patients with contagious disease are confined. Specialized protocols are used to decrease the risk that contaminated clothing and equipment will spread disease.

Laboratory   An area of the veterinary hospital in which testing occurs on biological samples such as blood, urine, feces, and cell aspirates.

Laboratory report   A report detailing the results and findings of any laboratory tests.

Large animal practice   A type of veterinary practice limited to the medical and surgical treatment of large animals—usually cattle, horses, pigs, sheep, goats, and similar species.

Legislative law   Laws that have been written, are approved, and then are enforced—such as speed limits and drinking ages.

Licensure   Legal permission granted to perform a set of tasks, usually by a board or agency that requires a defined level of competency.

Master Problem List   A portion of the problem-oriented medical record (POMR) that is a list of the patient's problems, the date they were first noted, the treatment plan and progress notes, and the date the problems were resolved.

Medical progress notes   A portion of the problem-oriented medical record (POMR), these notes describe in detail the ongoing medical and surgical management of an individual patient and the response to treatment.

Medical record   A written account of a patient's medical history, including diagnoses, treatments, response to therapy, prognosis, client communication, and any other documentation pertaining to patient care.

Mentor   A person who will supervise, encourage, and train those with less experience so they may develop confidence and competency.

Mirroring   The practice of reflecting a person's behavior in your own body language and gestures; engaging in similar actions to put another person at ease.

Mission statement   A written statement defining the purpose of a business or organization; e.g., "The mission of All Pets Animal Hospital is to provide competent medical and surgical care for pets by dedicated professionals."

Mixed animal practice   A type of veterinary practice that provides services to equine, food animal, and companion animal patients.

Mobile unit   A vehicle equipped to perform veterinary services at the site where animals are housed—most commonly a farm or stable, but may include house calls for companion animals.

Material Safety Data Sheets (MSDS)   Written documents supplied by the manufacturer describing the safe use of a product; how to appropriately manage, store, dispose of, and work with the substance in a safe manner.

National Association of Veterinary Technicians in America (NAVTA)   A national organization representing the profession of veterinary technology.

Necropsy   The procedure of examining a deceased animal to determine the cause of death.

Necropsy report   A report detailing the findings of a necropsy procedure.

National Institutes of Health (NIH)   An agency of the U.S. Department of Health and Human Services, the NIH consists of multiple institutes dedicated to improving health through research.

Nonverbal communication   Making one's feelings and attitudes known without speaking; presenting cues through body language to express one's thoughts.

Objective   A portion of the problem-oriented medical record (POMR) that indicates the observable and measurable observations of a patient's physical exam, such as mental state, temperature, weight, and heart rate.

One Health Initiative   A movement to encourage collaboration between all veterinary and human health care disciplines.

Occupational Safety and Health Administration (OSHA)   An agency of the U.S. Department of Labor, OSHA sets and enforces standards of workplace safety.

OSHA protocols   A set of written procedures designed to support workplace safety—e.g., a list of appropriate personal protective equipment required for working with diagnostic X-rays.

Paperless practice   A veterinary practice that utilizes software that stores data electronically, reducing or eliminating the need for paper medical records.

Paraphrasing   An indication of understanding given by rewording a client's statement and then asking for confirmation.

Pathology report   A written report documenting the cause of a disease, often leading to diagnosis. A pathology report is frequently requested to analyze tissue samples.

Patient information   The portion of the medical record regarding a pet's health or history.

Personality traits   Individual personal and social characteristics that shape behavior—e.g., kindness, empathy, respect.

Pharmacy   An area of the veterinary facility where drugs or other therapeutic agents are stored and dispensed.

Physical exam   A systematic process of observing an animal and assessing body systems; a combination of observation, palpation, and auscultation.

Plan   A portion of the problem-oriented medical record (POMR) that indicates what treatments and diagnostics are recommended to aid in therapy or diagnosis.

Problem-oriented medical record (POMR)   A method of organizing a medical record so that each individual health concern is recognized and consistently reevaluated for response to treatment.

Registry of Approved Continuing Education (RACE)   A program provided by the American Association of Veterinary State Boards (AAVSB) for the development and application of standards to providers of continuing education in veterinary medicine.

Radiology   A term used to describe the veterinary facility that houses a diagnostic radiology unit or other imaging equipment such as MRI or ultrasound. It refers to radiation used as therapy in the treatment of disease.

Referral practice   A type of veterinary practice where patients may be transferred for specialty procedures or surgeries. Chemotherapy, advanced diagnostics, and fracture repair are examples of common referral cases.

Referral report   A report from a referral practice describing in detail a patient's diagnostics, treatments, and prognosis.

Selective breeding   A process of choosing specific animals for breeding programs based on their desirable characteristics such as milk or egg production, color, size, or stature.

Signalment   The portion of the medical record that states the animal's age, breed/species, sex, and reproductive status.

Small animal practice   A type of veterinary practice typically limited to the treatment of dogs and cats.

SOAP   An acronym for the documentation of the problem-oriented medical record: subjective, objective, assessment, plan.

Somatic cell nuclear transfer   Also known as "cloning," a technique of selective breeding that uses genetic material from animals with desirable traits and is employed to reproduce genetically identical individuals.

Source-oriented medical record   A method of organizing a medical record by utilizing information from "the source" (the owner) and keeping this information in chronological order.

Subjective   A portion of the problem-oriented medical record (POMR) that is the description of events or clinical signs as reported by the owner; i.e., the owner's explanation of why the pet is at the veterinary hospital.

Surgery report   A portion of the medical record detailing all the events of a surgical procedure.

Surgery room   A room dedicated to performing surgical procedures.

Tact   A personality trait characterized by taking measures to avoid hurting someone's feelings.

Timed breeding   A method of synchronizing estrus and fertilization so that all animals are born within a limited time frame.

Transgenics   The transfer of DNA from one organism to another (see *genetically modified organism*).

Treatment area   An area of the veterinary facility designated, equipped, and supplied for medical treatments.

Trust   A personality trait characterized by being accountable and responsible for meeting your obligations.

U.S. Department of Agriculture (USDA)   Department of the federal government regulating agriculture, farming, and food safety through divisions such as the Animal and Plant Health Inspection Service (APHIS) and the Food Safety Inspection Service (FSIS).

Veterinary-Client-Patient Relationship (VCPR)   A relationship of mutual understanding between veterinarian and client that indicates that the veterinarian is familiar enough with the pet to make a diagnosis and recommendations for treatment. The veterinarian is also responsible for explaining any risks and expected outcomes, as well as where to seek emergency treatment.

Veterinary Medical Database   A collection of animal health information and statistics from colleges of veterinary medicine.

Veterinary Practice Act   A set of statutes devised to protect the health of animals by ensuring a set of standards and competencies for veterinary medical professionals.

Veterinary Technician National Exam (VTNE)   An exam administered by the American Association of Veterinary State Boards. It is used to evaluate the competency of an entry-level veterinary technician.

Zoonosis   Any disease that can be transmitted from animal to human.

Index

Note: Page numbers followed by 'b' indicates boxed material; those followed by "i" indicates an illustration; those followed by 'n' indicates a footnote; those followed by 't' indicates a table.